THE ORGANIZATIONAL MASTER PLAN HANDBOOK

A Catalyst for Performance Planning and Results

THE ORGANIZATIONAL MASTER PLAN HANDBOOK

A Catalyst for Performance Planning and Results

H. James Harrington • Frank Voehl

CRC Press
Taylor & Francis Group
Boca Raton London New York

CRC Press is an imprint of the
Taylor & Francis Group, an **informa** business
A PRODUCTIVITY PRESS BOOK

CRC Press
Taylor & Francis Group
6000 Broken Sound Parkway NW, Suite 300
Boca Raton, FL 33487-2742

Printed in the United States of America on acid-free paper
Version Date: 20120124

International Standard Book Number: 978-1-4398-7877-4 (Hardback)

Visit the Taylor & Francis Web site at
http://www.taylorandfrancis.com

and the CRC Press Web site at
http://www.crcpress.com

I dedicate this book to my wife, Marguerite, who has put up with me for over forty years. Her love makes living something to look forward to every day.

H. James Harrington

We dedicate this book to all those business leaders who wish to explore this brave new world of organizational alignment, with the help of experienced guides and counselors, in order to discover what works and what doesn't, and to apply the lessons learned to their own organizational and business networks.

H. James Harrington and Frank Voehl

Contents

Management Handbooks for Results

As series editors, we at CRC/Productivity Press have been privileged to contribute to the convergence of philosophy and the underlying principles of management for results, leading to a common set of assumptions. One of the most important deals with the challenges facing the transformation of the organization and suggests that managing for results can have a significant role in increasing and improving performance and strategic thinking, by drawing such experiences and insights from all parts of the organization and making them available to points of strategic management decision and action. As John Quincy Adams once said: "If your actions inspire others to dream more, learn more, do more, and become more, you are a leader."

If a good leader's actions inspire people to dream more, learn more, do more, and become more (and those actions will lead to an organization's culture, and if the culture represents "the way we do things around here"), then the Management for Results Series represents a brief glimpse of "the shadow of the leader-manager." The series is a compilation of conceptual management frameworks and literature review on the latest concepts in management thinking, especially in the areas of accelerating performance and achieving rapid and long-lasting results. It examines some of the more recent as well as historical contributions, and identifies a number of the key elements involved. Further analysis determines a number of situations that can improve the results-oriented thinking capability in managers, and the various handbooks consider whether organizations can successfully adopt their content and conclusions to develop their managers and improve the business.

This is a particularly exciting and turbulent time in the field of management, both domestically and globally, and change may be viewed as either an opportunity or a threat. As such, the principles and practices of management for results can aid in this transformation or (by flawed implementation approaches) can bring an organization to its knees. This Management for Results Series (and the handbooks contained therein) discusses the relationships among management thinking, results orientation, management planning, and emergent strategy, and suggests that management thinking needs to be compressed and accelerated, as it is essential in making these relationships more appropriate and effective—a so-called "shadow of the leader-manager." As series editors, we believe that the greater the sum total of management thinking and thinkers in the organization, the more readily and effectively it can respond to and take advantage of the vast array of changes occurring in today's business environment. However, despite the significant levels of de-layering and flattening of structures that has taken place in the last decade or so, some organizational barriers continue to stifle opportunities for accelerating management for results by limiting the flow of experiences and insights to relevant corners of the organization.

The "shadow of the leader-manager" that is present throughout this management handbook series is based upon the following eight characteristics of an effective leader-manager who gets results, and provides one of the many integration frameworks around which this Series is based:

1. Integrity = the integration of outward actions and inner values.
2. Dedication = spending whatever time or energy necessary to accomplish the task at hand, thereby leading by example and inspiring others.
3. Magnanimity = giving credit where it is due.
4. Humility = acknowledging they are no better or worse than other members of the team.
5. Openness = being able to listen to new ideas, even if they do not conform to the usual way of thinking.
6. Creativity = the ability to think differently, to get outside of the box that constrains solutions.
7. Fairness = dealing with others consistently and justly.
8. Assertiveness = clearly stating what is expected so that there will be no misunderstandings and dealing with poor performance.

This management book series is intended to help you take a step back and look at your team's or organization's culture to clearly see the reflection of your leader-manager style. The reflection you see may be a difficult thing for you to handle, but do not respond by trying to defend or to rationalize it as something not being of your making. As difficult as it may be, managers need to face the reality that their team's and organization's culture is a reflection of their leadership, leading to the concept of the leader-manager.

Accepting this responsibility is the first step to change, and as we all know, change begins with ourselves.

As Ghandi said many years ago, we all need to strive to become the "change we want to see in the world" .... in the case of management for results, we need to be the change we want to see in others!

Frank Voehl and H. James Harrington
Series Editors

Preface

> The Five Negative Ps—Poor Planning Produces Poor Performance.
>
> **—H. James Harrington**

Strategic planning has become both a manager's dream and nightmare. Virtually every type of organization in each and every industry around the world puts strategic planning alignment and accelerating results high up on its improvement agenda. Yet what is it about most organizations and their leaders that cause them to have such a hard time getting aligned behind their vision, mission, and strategic goals, and as a result, winding up getting rid of the employees they want to keep and keeping the ones they want to get rid of, or stifling the creativity and innovation they seek to foster? And why is it that a handful of companies are so successful at retention and innovation when others fail?

FedEx and UPS make it easy for customers to track their packages, create air bills, and schedule pickups on the web, but neighborhood banks cannot tell their customers the status of online bill payments made to the local phone company. In less than 10 years, Amazon went from a couple of employees in a warehouse to a more than $700 million business in 1999, all at the expense of the established bookstores Barnes & Noble and Borders. It takes only few minutes to purchase a flight, reserve a hotel room and a car through the Internet on Orbitz, but it takes five times as long to speak with an American or Delta travel agent.

While the logic supporting organizational alignment is compelling, why is it that Cisco, an internetworking company that makes routers and switches, can overhaul its product line every two years, but Kodak cannot seem to deliver rapid innovations to meet changing customer requirements? In short, what makes some companies successful in the new economy while others still continue to struggle to get it right? According to the research think tank e-Business Strategies, visionary companies understand that current business models and organizational designs are insufficient to meet the challenges of doing business in the new millennium. Enter the Organizational Master Plan methodology.

The Organizational Master Plan has allowed designers to make use of software as *components* instead of stovepipe applications and to draw applications together to achieve new accelerated levels of performance. The technical issues resulted from disparate infrastructures, systems that are begging to be integrated without a uniform standard for doing so, and processes that crossed departmental business boundaries, as if there was no common language for working together even if the systems were connected.

When special requirements are highlighted and further identified, these building blocks can be reorganized into any number of patterns to present a customized solution, without having to start from scratch each time. Large, or branded, organizations can use these technologies to move a supply chain network toward the desired level of connectivity, but smaller players can afford to participate in several networks simultaneously and participate as active and valuable contributors to their network performance.

I never make plans. You just have to change them.

—Failed CEO

For the visionary leaders, an Organizational Master Plan and its associated technology have become a key component for strategic decision making and learning, not only at the executive level, but also throughout the entire organization. In an aligned, performance-driven culture, strategy becomes everyone's everyday job; to challenge strategic assumptions, to assess risks, to evaluate the suitability of mergers and acquisitions, and most importantly, to facilitate decision making and actions. If you take a close look at such leading businesses as Dell, Cisco, Intel, Proctor & Gamble, Toyota, Walmart and Amazon, you'll find a new, uniquely aligned business design with some core technology—one that emphasizes a finely tuned integration of business, technology, goals, and process. In many cases, these companies are not only tapping technology to streamline operations, boost brands, improve customer loyalty, and ultimately drive profit growth, but visionary leaders are also setting new rules within their industries via new business models, aligned enterprise processes, and integrated operations to support changing customer requirements.

A good plan is a road map. It shows the final destination and usually the best way to get there.

—H. Stanley Judd, American Author

A major problem that management has been creating is the lack of a homogeneous continuous planning process. Typically, an organization follows this cycle: they do budgets in November, they do the Strategic Plan in January, and they address how the organizations need to improve in June. There has been lots and lots of planning, most of it done haphazardly, so the executive team can get back to doing something they think is more important, like answering customer complaints. The different plans were not linked together and the last one took priority over all of the previous ones. The executive team reasoned that things were changing so fast that planning was a waste of time. Most of the planning results were restricted to top management as they believed that as fast as things were changing, it would be confusing to communicate it to the line managers and the employees. It turned out that most of the planning was smoke and mirrors. Lots of the words that were being spoken were the *in* thing to be saying. Lots of excellent reports and work breakdown structures were being created, but with very little or no results.

To date, most of the planning has been directed at cost cutting by closing down factories and offices, or by moving the activity to another country. Information technology (IT) has become a very expensive crutch that often turns the organization's customers off. We stopped planning on "growing our people" and focused on the IT shop. A sweet-voiced customer service rep was replaced with an automated phone system. The following is a personal example of this IT focus:

Step 1. I dial the phone number. Computer voice answers saying, "This is Ace Airlines. This phone call may be monitored for quality reasons."

Step 2. Select from the following:

- Dial 1 if you want to know departure times.
- Dial 2 if you want to know arrival times.
- Dial 3 if you want to make a reservation.
- Dial 4 if you want to change your reservation.
- Dial 5 for advantage club information.
- Dial 6 for special bargains.
- Dial 7 for other subjects.

Step 3. I dialed 3. A computer voice comes on offering the following selection:

- Dial 1 if are going to travel outside the US.
- Dial 2 if you are going to travel within the US and if you are going to travel in the next 30 days.

- Dial 3 if you are going to travel within the US but not within the next 30 days.

Step 4. I dialed 1. A computer voice states, "Our reservation clerks are all busy due to the large amount of calls. You will be connected with a reservation clerk within the next 20 minutes."

Step 5. I sit there for 23 minutes listening to music, commercials, and the computer voice saying, "Please stay on the line. You are a very important customer to us."

Step 6. A computer voice came on the line saying, "Your call will be answered soon. Please dial in your frequent flyer number."

Step 7. I dialed in my frequent flyer number. A computer voice responded, "You are an Executive Club member. We give special service to our executive club members. Please hang up and dial 800-408-9000 to make your reservations.

Step 8. I was automatically disconnected.

Step 9. I repeated steps 1 through 8 because I didn't get the Executive Club phone number.

Step 10. I dialed 800-408-9000 and I repeated steps 1 through 7, only my wait time was only 10 minutes.

Step 11. This time I reach a woman with a strong accent who I had a hard time understanding. After a considerable amount of effort, I did get my reservations. She explained that there would be a $100 service charge because I didn't use the Internet. She also explained a service charge of $400 and 300 dirhams if I wanted to change any of my flight segments. She went on to inform me that for better service, I should use the Internet next time.

(*Note:* I was flying from San Francisco, California, to Dubai, United Arab Emirates, in business class, which is more than a $10,000 fare. You have to question if the airline wants to jeopardize a customer's business just to save the cost of a reservation clerk's time. Actually, it appears that this airline company wants me to buy a computer and learn how to use it, just so that they can save a little money. Is it any wonder that customer satisfaction is spiraling downward in more organizations that are using these automated systems to save on personnel costs?)

> Planning is a process of choosing among those many options. If we do not choose to plan, then we choose to have others plan for us.
>
> **—Richard L. Winwood, American Attorney**

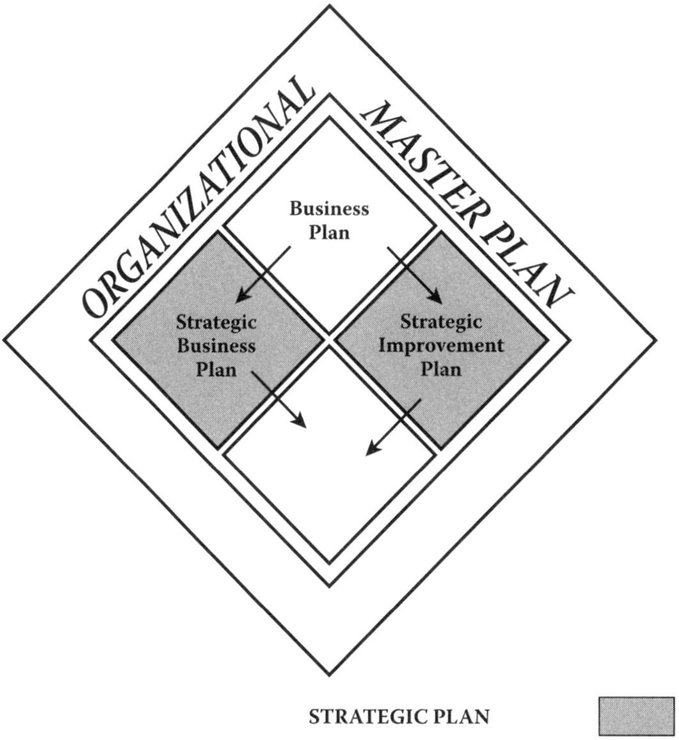

FIGURE 0.1
The five parts of the Organizational Master Plan.

The Organizational Master Plan must combine all of the major planning processes together, and at the same time ensure that they are customer focused. All of the plans must be based on understanding customer requirements and preferences, plus having an excellent understanding of what the competition is doing currently and will be doing in the future. Figure 0.1 is a model of an Organizational Master Plan, indicating the different parts of this plan: the Business Plan, the Strategic Business Plan, the Strategic Improvement Plan, and the Strategic Plan. (The Strategic Plan is the prioritized combination of the Strategic Business Plan and the Strategic Improvement Plan; in the model it is represented by the combination of the two gray diamonds.)

The Business Plan's primary objective is to inform external investors. The Strategic Plan, which includes the Strategic Business Plan (SBP) and the Strategic Improvement Plan (SIP), focus on the future of the organization. And the focus of the Annual Operating Plan (AOP) is short range and addresses the current operations.

Acknowledgments

We want to acknowledge Candy Rogers, who converted and edited endless hours of dictation and misspelled words into the finished product. We couldn't have done it without her dedication, help, and proofreading.

We also would like to acknowledge the members of the senior management and boards of directors (BODs) of the many organizations with which we have worked over the past four decades, who have helped to form the frameworks, concepts, ideas, and methods outlined in this book. We would also like to thank the many sponsors who have helped us drive the research forward by participating in meetings, audio and video conferences, and roundtable meetings.

In addition, we would like to thank the American Society for Quality (ASQ) and the International Academy for Quality for their unfailing support for advancing many of these concepts on the national and international level.

H. James Harrington and Frank Voehl

About the Authors

H. JAMES HARRINGTON

Dr. H. James Harrington is one of the world's quality system gurus with more than sixty years of experience. In the book, *Tech Trending* (Amy Zuckerman, Capstone, 2001), Dr. Harrington was referred to as "the quintessential tech trender." The *New York Times* referred to him as having a "knack for synthesis and an open mind about packaging his knowledge and experience in new ways—characteristics that may matter more as prerequisites for new-economy success than technical wizardry." He has been involved in developing quality management systems in Europe, South America, North America, Middle East, Africa, and Asia.

Dr. H. James Harrington now serves as the chief executive officer for the Harrington Institute. He also serves as the chairman of the board for a number of businesses and as the US chairman of Chair on Technologies for Project Management at the University of Quebec in Montreal. Dr. Harrington is recognized as one of the world leaders in applying performance improvement methodologies to business processes.

In February 2002, Dr. Harrington retired as the chief operating officer (COO) of Systemcorp A.L.G., the leading supplier of knowledge management and project management software solutions. Prior to this, he served as a principal and one of the leaders in the Process Innovation Group at Ernst & Young. Dr. Harrington was with IBM for over thirty years as a senior engineer and project manager.

Dr. Harrington is past chairman and past president of the prestigious International Academy for Quality and of the American Society for Quality Control. He is also an active member of the Global Knowledge Economics Council.

The Harrington/Ishikawa Medal presented yearly by the Asia-Pacific Quality Organization was named after Dr. Harrington to recognize his many contributions to the region. In 1997, the Quebec Society for Quality named their Quality Award "The Harrington/Neron Medal," honoring Dr. Harrington for his many contributions to the Quality Movement in Canada. In 2000 the Sri Lanka national quality award was named after him. The Middle East and Europe Best Quality Thesis Award was named "The Harrington Best TQM Thesis Award." The University of Sudan has established a Harrington Excellence Chair to study methodologies to improve organizational performance. The Chinese government presented him with the Magnolia Award for his major contribution to improving the quality of Chinese products.

Dr. Harrington's contributions to performance improvement around the world have brought him many honors and awards, including the Edwards Medal, the Lancaster Medal, the American Society for Quality's Distinguished Service Medal, and many others. He was appointed the honorary advisor to the China Quality Control Association, and was elected to the Singapore Productivity Hall of Fame in 1990. He has been named lifetime honorary president of the Asia Pacific Quality Organization and honorary director of the Association Chilean de Control de Calidad.

Dr. Harrington has been elected a Fellow of the British Quality Control Organization and the American Society for Quality Control. He was also elected as an honorary member of the quality societies in Taiwan,

Argentina, Brazil, Colombia, and Singapore. He is also listed in *Who's Who Worldwide* and *Men of Distinction Worldwide.* He has presented hundreds of papers on performance improvement and organizational management structure at the local, state, national, and international levels.

Dr. Harrington is a very prolific author, publishing hundreds of technical reports and magazine articles. He has authored 33 books and 10 software packages. His e-mail address is hjh@harrington-institute.com.

FRANK VOEHL

Frank Voehl now serves as the chairman and president of Strategy Associates, Inc. and as a senior consultant and chancellor for the Harrington Institute. He also serves as the chairman of the board for a number of businesses and as a Master Black Belt Instructor and Technology Advisor at the University of Central Florida in Orlando, Florida. He is recognized as one of the world leaders in applying quality measurement and Lean Six Sigma methodologies to business processes.

Frank Voehl has extensive knowledge of National Regulatory Commission (NRC), Food and Drug Administration (FDA), Good Manufacturing Practice (GMP), and National Aeronautics and Space Administration (NASA) quality system requirements. He is an expert in ISO-9000, QS-9000, ISO-14000, and Six Sigma Quality System Standards and processes. He has degrees from St. John's University and advanced studies from New York University, as well as a Doctor of Divinity degree. Since 1986, he has been responsible for overseeing the implementation of quality management systems with organizations in such diverse industries as telecommunications and utilities; federal, state, and local government agencies; public administration and safety; pharmaceuticals; insurance and banking; manufacturing; and institutes of higher learning. In 2002, he joined The Harrington Group as the chief operating officer and executive vice president. Has held executive management positions with Florida Power and Light and FPL Group, where he was the founding general manager, and chief operating officer of QualTec Quality Services for seven years. He has written and published or copublished over twenty-five books and hundreds of technical papers on business management, quality improvement, logistics and team building, and has received numerous awards for community leadership, service to the third world countries, and student mentoring.

The Bahamas National Quality Award was developed in 1991 by Voehl to recognize the many contributions of companies in the Caribbean region, and he is an honorary member of its Board of Judges. In 1980, the City of Yonkers, New York, declared March 7 as Frank Voehl Day, honoring him for his many contributions on behalf of the youth in the city where he lived and performed volunteer work. In 1985 he was named Father of the Year in Broward County, Florida. He also serves as president of the Broward County St. Vincent de Paul Society, whose mission is to serve the poor and needy.

Frank's contributions to quality improvement around the world have brought him many honors and awards, including the American Society for Quality Distinguished Service Medal, the Caribbean Center for Excellence Founders Award, the Community Quality Distinguished Service Award, the Czech Republic Outstanding Service Award on behalf of its business community leaders, FPL's Pioneer Lead Facilitator Award, the Florida SFMA Partners in Productivity Award, and many others. He was appointed the honorary advisor to the Bahamas Quality Control Association, and was elected to the Eastern Europe Quality Hall of Fame. He has been named honorary director of the Association Venezula de Control de Calidad by Banco Consolidado.

Other Books by H. James Harrington and Frank Voehl

BOOKS BY H. JAMES HARRINGTON

H. James Harrington is a very prolific author, publishing hundreds of technical reports and magazine articles. H. James Harrington's books have been published by McGraw-Hill, ASQ's Quality Press, Paton Press and CRC Press, have been published in Chinese, Russian, Italian, Spanish, Arabic, Portuguese, French, Romanian, Hebrew, Swedish, and Korean. He has authored 35 books, which include

- *The Improvement Process* (McGraw-Hill 1987)
- *Poor-Quality Cost* (Marcel Dekker 1987)
- *Excellence—The IBM Way* (American Society for Quality 1988)
- *The Quality/Profit Connection* (ASQ Quality Press 1988)
- *Business Process Improvement* (McGraw-Hill 1991)
- *The Mouse Story* (Ernst & Young 1991)
- *Of Tails and Teams* (ASQC Quality Press 1994)
- *Total Improvement Management* (McGraw-Hill 1994)
- *High Performance Benchmarking* (McGraw-Hill 1995)
- *The Complete Benchmarking Workbook* (McGraw-Hill 1996)
- *ISO 9000 and Beyond* (McGraw-Hill 1996)
- *The Business Process Improvement Workbook* (McGraw-Hill 1997)
- *The Creativity Toolkit—Provoking Creativity in Individuals and Organizations* (McGraw-Hill 1998)
- *Statistical Analysis Simplified—The Easy-to-Understand Guide to SPC and Data Analysis* (McGraw-Hill 1998)

- *Area Activity Analysis—Aligning Work Activities and Measurements to Enhance Business Performance* (McGraw-Hill 1998)
- *ISO 9000 Quality Management System Design: Optimal Design Rules for Documentation, Implementation, and System Effectiveness* (ISO 9000 Quality Management System Design) (coauthor; ASQ Quality Press 1998)
- *Reliability Simplified—Going Beyond Quality to Keep Customers for Life* (McGraw-Hill 1999)
- *ISO 14000 Implementation—Upgrading Your EMS Effectively* (McGraw-Hill 1999)
- *Performance Improvement Methods—Fighting the War on Waste* (McGraw-Hill 1999)
- *Project Change Management—Applying Change Management to Improvement Projects* (McGraw-Hill 1999)
- *Simulation Modeling Methods—An Interactive Guide to Results-Based Decision Making* (McGraw-Hill 2000)
- *E-Business Project Manager* (Paton Press 2002)
- *Process Management Excellence—the Art of Excelling in Process Management* (Paton Press 2005)
- *Project Management Excellence—the Art of Excelling in Project Management* (Paton Press 2005)
- *Change Management Excellence—The Art of Excelling in Change Management* (Paton Press 2005)
- *Resource Management Excellence—The Art of Excelling in Resource Management* (Paton Press 2005)
- *Six Sigma Statistics Simplified*; in Arabic (2006)
- *Improving Healthcare Quality and Cost with Six Sigma* (FT Press 2007)
- *Making Teams Hum* (Mandarin) (Standards Press of China 2007)
- *Advanced Performance Improvement Approaches: Waging the War on Waste II* (e-TQM College 2007)
- *Fast Action Solution Technique* (Paton Press 2008)
- *Corporate Governance: From Small to Mid-Sized Organizations* (Paton Press 2009)
- *Streamlined Process Improvement* (McGraw-Hill 2011)

BOOKS BY FRANK VOEHL

Frank Voehl has authored, co-authored and is the series editor of over 30 books. Voehl's books have been published by CRC Press, St. Lucie Press, or ASQ's Quality Press and Strategy Publications Center (Self-Published), and have been published in English, Italian, Spanish, Arabic, and Portuguese. They include the following:

- *Deming: The Way We Knew Him* (St. Lucie Press 1993)
- *ISO 9000: An Implementation Guide for Small to Mid-Sized Businesses* (St. Lucie Press 1993)
- *The House of Quality: On the Five Theorists and the Minor Prophets* (Strategy Publications Center 1993)
- *Total Quality: Principles and Processes Within Organizations* (St. Lucie Press 1993)
- *Total Quality Series: TQM Implementation Handbook* (St. Lucie Press 1993)
- *Total Quality Series: Information Systems/Technology* (St. Lucie Press 1994)
- *Total Quality Series: Purchasing/Supply Management* (St. Lucie Press 1994)
- *Total Quality Series: TQ in Higher Education* (St. Lucie Press 1994)
- *Total Quality Series: TQ in Human Resource Management* (St. Lucie Press 1994)
- *Total Quality Series: TQ in Radiology* (St. Lucie Press 1994)
- *Total Quality Series: TQ in Organizational Development* (St. Lucie Press 1995)
- *Teambuilding: A Structured Learning Approach* (St. Lucie Press 1996)
- *Teambuilding: A Structured Learning Approach (Instructor Guide)* (St. Lucie Press 1996)
- *The Executive Guide to Quality Systems* (St. Lucie Press 1996)
- *The Executive Guide to Quality Systems (Instructor Guide)* (St. Lucie Press 1996)
- *Total Quality Series: TQ in Marketing* (St. Lucie Press 1996)
- *Total Quality Series: TQ in Research and Development* (St. Lucie Press 1996)
- *APICS Resource Management Series: Applying Manufacturing Execution Systems* (CRC Press 1997)

- *APICS Resource Management Series: Macrologistics Management* (CRC Press 1998)
- *APICS Resource Management Series: Supply Chain Management* (CRC Press 1998)
- *For Results Series: Facilitation for Results* (Strategy Publications Center 2000)
- *For Results Series: Problem Solving for Results* (St. Lucie Press 1998)
- *For Results Series: Visioning for Results* (Strategy Publications Center 2001)
- *ISO-TAP: A Transition of the standard from 1994 to 2000* (Strategy Publications Center 2001)
- *ASQ Recovering Prosperity Through Quality: CSBMR/DD Case Study* (ASQ Press 2002)
- *For Results Series: SPC for Results* (Strategy Publications Center 2002)
- *Recovering Prosperity Through Quality: Community Improvement Handbook* (ASQ Press 2003)
- *Lean Six Sigma (LSS) Playbook* (Strategy Publications Center 2008)

BOOKS BY H. JAMES HARRINGTON AND FRANK VOEHL

- *Knowledge Management Excellence—The Art of Excelling in Knowledge Management* (Paton Press 2005)
- *Six Sigma Green Belt Workbook* (Paton Press 2008)
- *Six Sigma Yellow Belt Workbook* (Paton Press 2008)
- *The Organizational Alignment Handbook: A Catalyst for Performance Acceleration* (CRC Press 2011)
- *Organizational Master Plan* (CRC Press 2012)

1

Overview of the Problem

> When only top management own the plan, they have to spend most of their time pushing it or it won't get done.
>
> —**H. James Harrington**

World War III has begun. This time it is not a war of battleships, bullets, and bombs—this is an economic war. The weapons are televisions, steel, cars, and clothes. This is a war where we have no allies. Every country in the world is out to capture more of its share of the US and world markets. We are being attacked with tires from Brazil, cars from Japan, radios from Taiwan, clothes from China, cosmetics from France, shoes from Italy, beef from Argentina and Australia, and so on.

US business entered the 1980s with a deep-seated resolution to stop the flood of import products and as a result, a group of new admirals and generals took over to reestablish our industrial leadership. These were people like John Akers of IBM, F. James Mc Donald of General Motors, Jim Olson of AT&T, and John Young of Hewlett Packard. Industrial leaders like these laid out strategies to provide a thrust to lead the United States back to the prominence it once had, but it takes years to reestablish a reputation once it has been destroyed or at least tarnished.

> Think ahead. Don't let day-to-day operations drive out planning.
>
> —**Donald Rumsfeld (former US Secretary of Defense)**

General Motors (GM)—one of the most powerful and respected organizations—filed for bankruptcy in 2009. Why did this happen? What did they try to do that didn't work? To help understand this, I will report on an interview I had in 1988 with GM's corporate president, F. James McDonald, which was documented in Dr. Harrington's book *The Quality/Profit Connection* (published by Quality Press in 1989).

General Motors celebrated its 100-year anniversary on September 16, 2008. It was on this date in 1909 that William C. Durant founded General Motors Company, predecessor of the current General Motors Corporation. Durant's first acquisition for General Motors Corporation was the Buick Motor Company.

In 1988, GM had 151 facilities operating in 26 states and 90 cities in the United States and 13 plants in Canada. It had assembly, manufacturing, distribution, sales, or warehousing operations in 37 other countries. GM also had equity interest in associated companies that conducted assembly, manufacturing, or distribution in several countries. The average worldwide employment totaled approximately 748,000 men and women in 1984.

In January 2009 GM claimed to employ 244,500 people globally, which was a drop of 503,500 employees from their 1984 head count. Just 6 months later, that number fell by at least 20,000 as GM announced the closing of 14 plants and three warehouses. On the dealership front, between GM and Chrysler, nearly 3,000 franchises have closed their doors. GM had dropped from the Number 1 seller of cars to Number 2 and was moving fast down to becoming Number 3. The US government stepped in and invested millions of taxpayer dollars to try to keep GM operational to no avail. Now GM, once the pride of the U.S. companies, has gone bankrupt.

Here is an excerpt from Dr. Harrington's interview with F. James McDonald.

Harrington: What were the circumstances leading to the current focus of GM on quality improvement?

McDonald: Efficient, small, high-quality vehicles from Japan and the availability of these vehicles at just the right time in history were watershed events in the US auto industry. Their perceived quality became the benchmark for all cars—in effect, customer standards changed dramatically. And that change swept through the entire line of products.

Harrington: Do you have an official quality policy?

McDonald: Actually, the new quality consciousness at GM began with the development of a quality ethic for all GM units and operations. The essence of this ethic boils down to this: Quality is the number one operating priority at GM today.

Harrington: To what sections of the business is it being applied?

McDonald: Quality improvement is being applied to all areas of our business. Specific quality objectives and strategies must be included within each unit's five-year Business Plan. All departments within a business, and of course each employee, contribute to meeting those quality objectives.

On new product programs, resources are allocated very early when our ability to influence the outcome is greatest. This includes the front loading of people from all disciplines including marketing, product engineering, manufacturing, assembly, quality assurance financial, and materials management. This includes early sourcing decisions so our suppliers can work with product development teams on potential problems and improvement.

Harrington: What activities were undertaken to start the quality improvement process and when did it start?

McDonald: At GM today, we have this kind of strategic vision, and that vision is simply to offer world-class quality in every market segment. By world class, we mean parity with, or superiority to, the best in the field—product for product.

To assist the operating units in this effort, the corporation has issued four key success factors for quality, which help focus the GM quality ethic and its six mandates. Research has shown that these key success factors must be addressed in business planning and implementation strategies if meaningful quality improvement is to occur.

Let's take a look at what the key success factors and the associated objectives are.

- Management commitment. Managers at all levels must be committed to continuous quality improvement and demonstrate their commitment by word or action.
- People-development process. Every employee, regardless of function or level, must have the encouragement, support, and opportunity to be a contributing member of the quality improvement effort.
- Quality performance processes. Each task and activity must have processes and tools to ensure conformance to specifications and to provide for continuous quality improvement.

- Customer satisfaction. General Motors must be the world leader in quality, reliability, durability, performance, service, and value, as confirmed by customer-defined measures and marketplace response.

We have also identified the major activities to carry out improvements on any given project and have found that they work quite well.

Harrington: What is the role of top management in the improvement process?

McDonald: Achieving true quality maturity is totally the responsibility of top management in our company. Others may carry it out to one degree or another, but those at the top must be willing to go the whole route.

We believe that the whole top management team must be aboard. Even the most inspiring leader can't hope to reach the organization without total commitment from everyone at the top.

Harrington: What is the role of the employees and the union in the improvement process?

McDonald: We are absolutely convinced that eventual success depends heavily on the employees. As we discussed, one of our key success factors for quality improvement concerns people-development processes.

For instance, we've trained more than 30,000 GM workers in statistical process control (SPC) techniques. And I must say, to see these tools put to work right on the line is one of the most rewarding experiences I've had at GM. So, I think we're on the right track on the employee side—even though we still have a ways to go.

Harrington: What problems did you have in implementing the improvement process?

McDonald: We're now in a sort of middle ground on the quality cycle. Currently, techniques such as SPC at the point of manufacture are providing outstanding improvement, as verified by the more recent new products. Statistical techniques are helping to ensure that the process is being controlled. The whole focus must be on process control, not inspection of the parts.

But prevention within manufacturing can take you only so far along the journey. Greater success must come from moving the focus upstream to design and engineering, for example, by combining the talents of design engineering, processing,

> and manufacturing, and having them work together as a team instead of individually. That's the place to start if you're serious about doing everything right the first time. Our product development teams on new products that we have previously mentioned are addressing this in a fine manner. We are also initiating this concept in our daily operations.
>
> GM's reorganization of its North American passenger cars and its worldwide truck and bus operations addressed changes necessary to ensure quality improvements, accountability for results, and effective allocation of resources. The reorganization was quality driven from the beginning.

On reviewing McDonald's comments, Dr. Harrington acknowledges that he was saying all the right things and doing all the right things, but what went wrong? Are we telling management to do the wrong things? Is the quality message one that leads an organization into bankruptcy?

We believe with all our hearts that what McDonald was saying were the right things:

1. He recognized that the playing field had changed and the organization had new benchmarks.
2. He stated that quality is the number one operating priority at GM.
3. He stated that quality objectives and strategies are part of their five-year Business Plan.
4. He had GM doing concurrent design with early supplier involvement.
5. They had a strategic vision and that vision was simply to offer world-class quality in every market segment. By "world-class," he explained that they must be equal to or better than the best in the field, product for product.
6. They had four key success factors:
   - Management commitment
   - People development processes
   - Quality performance processes
   - Customer satisfaction
7. He stated that GM must be the world leader in quality, reliability, durability, performance, service, and value as confirmed by customers as confirmed by customer-defined measurements and marketplace response.

8. He made the top managers in each organization totally responsible for achieving the objectives and implementing the strategies.
9. He had over 30,000 employees trained related to performance improvement.
10. He recognized that GM needed to stop being the best at problem solving and start being the best at prevention of problems.
11. He required management at all levels to be committed to the plan and demonstrate their commitment by word and action.
12. He had made North American passenger cars and GM's worldwide trucks and bus operations address changes necessary to improve quality improvement, accountability for results, and effective allocation of resources.

Saying and planning to do the right things is admirable, but as GM has proved, it is the *way* you do things—how management behaves, how they spend their time, what they are interested in—that makes the difference between success and failure. Was McDonald really serious when he said "GM must be the leader in quality, reliability, durability, performance, service, and value as confirmed by customer-defined measurements and market response"? Did heads roll because GM has never fixed its reliability problems?

Dr. Harrington doesn't believe that the GM plans had the driving force behind them that they needed to be successful. He still remembers a full-page ad in *USA Today*, paid for by GM, where it showed a graph indicating the quality level of GM cars compared to Japanese cars. It showed that over the years, GM quality was worse than Japanese cars, and that at the present time, GM had improved to the point that its quality was only 20% worse than Japanese cars. Wasn't that just like telling the GM customers and everyone else not to buy GM, but buy Japanese cars in their place?

To be successful in today's environment, you have to have a combined-focus planning process that brings the total organization together, marching like soldiers down the highway in the Fourth of July parade. Everyone needs to be in step with the tune the band is playing. We can no longer afford to have the band playing three or four different tunes at the same time. And the president (chief executive officer [CEO]) needs to be out in front like the band leader where everyone's eyes are on him/her and where the slightest deviation from the plan is immediately recognized. Along behind the president is the top management team who serves like the baton twirlers in the parade. They follow the president, but they are in perfect harmony with him/her. This is what a Master Plan will do for your organization, but it can't do it by itself. The Organizational Master Plan is the starting point, not the objective; it is the first step in getting the organization working in the same direction and in harmony. Where the "rubber meets the road" is when the plan is implemented and the results are measured. McDonald had GM's vision well defined and a strategy well developed to get them there. The question remains, "How well was it implemented over the past twenty years? Remember, part of the vision was "GM must be the world leader in quality, reliability, durability, performance, service, and value as confirmed by customer-defined measurements and marketplace response." Well, the quality of GM cars is much better today than it was in the past, but they are still not the world leader. Toyota is. Reliability, durability, performance, and as a result value, is poor as compared to Ford and many Asian cars. Dr. Harrington has sat in many presentations made by GM people at conferences. Most of them were about what they were doing to improve supplier products. None of them addressed what GM was doing to improve the quality and reliability of their design operations. GM went bankrupt not because of lack of a Strategic Plan, but because its Master Plan didn't effectively implement the Strategic Plan across all the units within GM. In other words, their Organizational Master Plan did not accomplish the desired results year after year for the last twenty years.

What did twenty years of improvement do for GM?

- They dropped from the Number 1 car producer, Toyota took over, and it looks like they will be Number 3 soon.
- At the start of 2009 they had 244,500 employees globally. That is a drop of 503,500 employees from their 1984 high of 748,000. GM dropped 20,000 more employees in June 2009.
- They announced the closing of 14 plants and 3 warehouses.
- They sold Hummer to a Chinese manufacturer.
- Saturn, the car they designed to compete with the low-cost high-quality Japanese cars, was sold to Roger Penske.
- They sold Pontiac to Kaput.
- They plan on selling Opel and Vauxhall in Europe.
- They went into a joint venture with Toyota so they could learn Toyota's manufacturing approach and set up the Nummi plant in Fremont, California, only to drop out of it and put the jobs of almost 3,000 people in jeopardy.
- They put at least 40 of their major suppliers into bankruptcy.
- They closed almost 3,000 dealerships.

It was twenty years of poor planning and implementation that caused GM to keep slipping downhill to get into this terrible state of affairs.

> Planning is an unnatural process; it is much more fun to do something. The nicest thing about not planning is that failure comes as a complete surprise, rather than being preceded by a period of worry and depression.
>
> **—Sir John Harvey-Jones**

The purpose of this book is to show the reader how to design and manage an Organizational Master Plan, which brings together the five different parts of this plan: the Business Plan, the Strategic Business Plan, the Strategic Improvement Plan, the Strategic Plan, and the Annual Operating Plan. (The Strategic Plan is the prioritized combination of the Strategic Business Plan and the Strategic Improvement Plan; in the model it is represented by the combination of the two gray diamonds.) Figure 1.1 is a view of the parts and interfaces of the Organizational Master Plan.

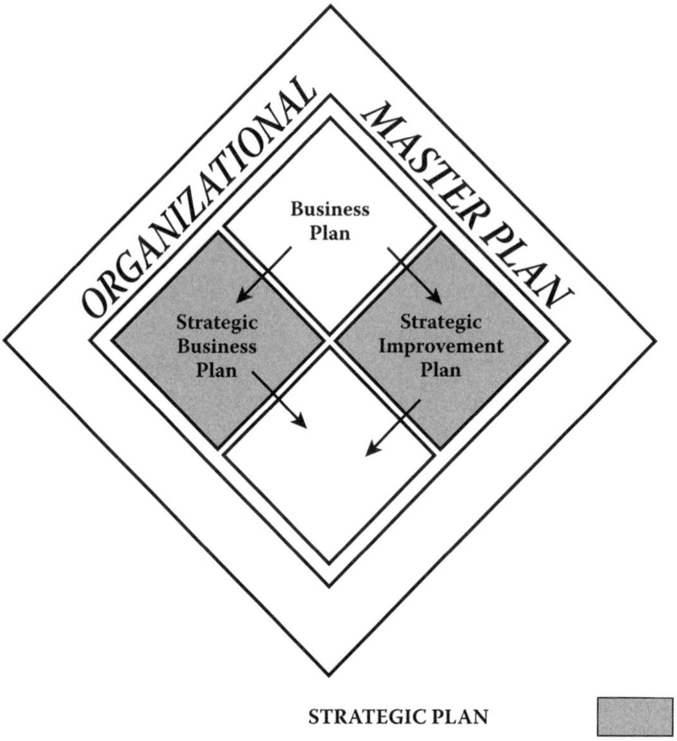

FIGURE 1.1
The Organizational Master Plan.

ORGANIZATIONAL PLANS

The following are definitions of the key business planning terms used throughout this book.

The Annual Operating Plan: The Annual Operating Plan (AOP) is a formal statement of business short-range goals, the reasons they are believed to be attainable, the plans for reaching these goals, and the funding approved for each part of the organization (budget). It includes the implementation plan for the coming years (years 1–3) of the Strategic Plan. It may also contain background information about the organization or teams attempting to reach these goals. One of the end results is performance plans for each manager and employee who will be implementing the plan over the coming year. The AOP is often just referred to as the Operating Plan (OP).

Business Plan: The Business Plan is a formal statement of a set of business goals, the reasons why they are to be believed attainable, and the plan for reaching those goals. It also contains background information about the organization or team attempting to reach those goals.

Organizational Master Plan: The Organizational Master Plan is the combination and alignment of an organization's Business Plan, Strategic Business Plan, Strategic Improvement Plan, Strategic Plan, and Annual Operating Plan.

The Strategic Plan: The Strategic Plan is a document that is the result of strategic planning. It defines the organization's strategy and direction and makes decisions on the allocation of resources in pursuit of the organization's strategy including its capital and people. It focuses on the future of the organization and is the combination of the Strategic Business Plan and the Strategic Improvement Plan with each item prioritized to maximize the organization's performance.

The Strategic Business Plan: The Strategic Business Plan (SBP) focuses on what the organization is going to do to grow its market. It is designed to answer the questions: What do we do? How can we beat or avoid the competition? It is directed at the product and/or services that the organization provides as viewed by the outside world. When an organization is just being funded, this plan is often just called a Business Plan.

The Strategic Improvement Plan: The Strategic Improvement Plan (SIP) focuses on how to change the culture of the organization. It is designed to answer the questions: How do we excel? How can we increase value to all the stakeholders? It addresses how the controllable factors within the organization can be changed to improve the organization's reputation and performance.

Strategic Management: Strategic Management (SM) is the process of specifying the organization's mission, vision, and objectives, developing policies and plans, often in terms of projects and programs that are designed to achieve these objectives and then allocating resources to implementation of policies and plans, projects, and programs.

TERMS

Throughout this book, the following terms will be used:

Department: A department is made up of a manager and the people who report directly to him or her. It covers all levels of management—top, middle, and line.

Executive team: An executive team is a management team, which usually includes the chief executive officer (CEO), chief operating officer (COO), and vice presidents of an organization. In some organizations, senior key staff people are also included.

Natural work teams: Natural work teams are groups of people that consist of a manager or a team leader and the people who report directly to him or her.

Function: A function is an organization that is made up of a specific skill set of people (i.e., Quality Assurance, Finance, Human Relations, Production Control, Sales, marketing, Product Engineering, Manufacturing, Industrial Engineering, etc.) or the total group of people that reports directly to a vice president of the organization.

Wishing consumes as much energy as planning.

—Anonymous

PARTS OF THE ORGANIZATIONAL MASTER PLAN

The Organizational Master Plan is divided into five different but related types of plans that overlap in some areas. They are

- **Business Plan:** This plan's primary use is to inform potential investors about the organization. It provides input into the Strategic Business Plan and the Strategic Improvement plan.
- **Strategic Business Plan and the Strategic Improvement Plan:** Together, these two plans make up the Strategic Plan. Its focus is on the future. These plans are used to set direction, not to tell the organization how to do it. The Strategic Business Plan and the Strategic Improvement Plan should go through a major update every 3 to 5 years and be reviewed every year.
- **Strategic Plan:** The Strategic Plan combines the Strategic Business Plan and the Strategic Improvement Plan so that the individual activities are in harmony and prioritized based upon value-added to the organization.
- **Annual Operating Plan:** This plan's focus is on the current or present short-range activities. It covers the tasks that are scheduled to be done and when they will be done. The Annual Operating Plan should be redone at least once a year and reviewed once a month and whenever a major tollgate is scheduled.

Yes, the Organizational Master Plan is the heart of the transformation. But the heart needs veins to pump life-giving blood throughout the organization. Without this, the heart is of no value, and when one of these major veins gets clogged, the body stops functioning. This is true of the Organizational Master Plan. It serves as the pump that provides the direction and it is the way that direction is implemented that makes the organization a success or a failure.

There are many ways to implement the Organizational Master Plan, but the two best approaches are organizational alignment (OA) and policy deployment (PD).

Organizational alignment: Organizational alignment is the methodology that brings the organization's structure, processes, networks, people, and reward system in harmony with the Strategic Business Plan and the Strategic Improvement Plan.

Policy deployment: Policy deployment (also called *hoshin kanri*) is a method devised to capture and cement strategic goals as well as flashes of insight about the future and develop a means to bring these into reality. It is based upon Shewhart's plan-do-check-act cycle. It creates goals, selects control points, and links daily control activities to the organization's strategy.

THE ORGANIZATIONAL ALIGNMENT CYCLE

Organizations that have been successful, based in part on good planning, know that they must provide the organization and its employees with a road map to help "translate" the vision and mission into *things people can do.* The next most critical element of what's in a good plan is the strategic focus on the organization in terms of *how* it will compete.

- **Mission:** The mission is the stated reason for the existence of the organization. It is usually prepared by the chief executive officer and seldom changes, normally only when the organization decides to pursue a completely new market.
- **Vision:** The vision is a documented or mental description or picture of a desired future state of an organization, process, team, or activity. It is usually prepared by top management and it is directed at what the organization's output will be like and how it will be used 10–25 years in the future. Some organizations have a long-range vision statement (10–25 years) and a short-range vision statement (5–10 years).
- **Value statement:** Value statements are documented directives that set behavioral patterns for all employees. They are the basic beliefs that the organization is founded upon, the principles that make up the organization's culture. They are deeply engrained operating rules or guiding principles of an organization that should not be compromised. Value statements are sometimes called operating principles, guiding principles, or basic beliefs' operating rules.
- **Strategic focus:** These are the key things that will set the organization apart from its competitors over the next 5 years. This list is defined by top and middle-level managers.

The Organizational Alignment Cycle (OAC) basically says that an organization should align six elements to achieve a sound organizational design: strategy, processes, structure, people, rewards, and implementation as illustrated in Figure 1.2.

The OAC is divided into six phases:

- **Phase I. Strategic Plan:** This phase defines directions and sets expectations. It specifies the goals and objectives to be achieved, as well as

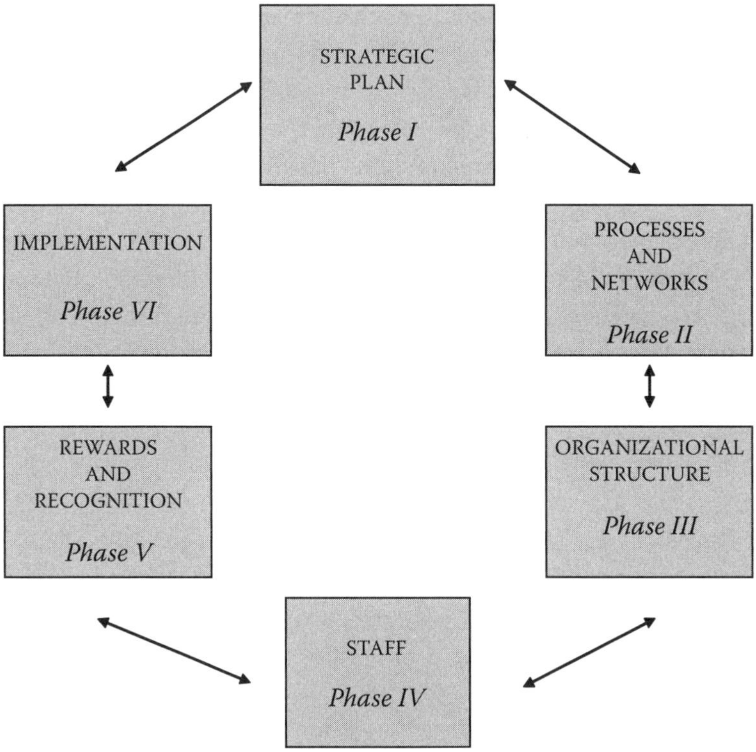

FIGURE 1.2
The Organizational Alignment Cycle.

the values and missions to be pursued; it sets out the basic *direction* of the organization. The Strategic Plan includes these two different plans:
- Strategic Business Plan
- Strategic Improvement Plan

- **Phase II. Processes and Networks Design:** This phase defines the way activities are joined together and the way information is distributed. It outlines the functioning of an organization and the circulation of *information*, and how outputs are generated.
- **Phase III. Organizational Structure Design:** This phase defines the responsibilities and decision-making power within the organization. Through the organizational and the departmental design, it determines the placement of *power and authority* in the organization.
- **Phase IV. Staffing:** This phase defines the type of people who are needed and the skills they must have. It outlines the skills and mindsets required by the strategy and structure of the organization.

- **Phase V. Rewards and Recognition System Design:** This phase is used to motivate and encourage people to behave in the desired behavior pattern. It aligns the goals of the employee with the goals of the organization by providing motivation and incentives for the completion of the organization's strategic direction.
- **Phase VI. Implementation:** This phase includes the processes and activities that are required to bring about the desired changes in the organization. It includes preparing the procedures and policies that allow the organization to operate at maximum efficiency. It defines how the changes are installed into the organization.

> Organizational alignment is a process, not six disjointed different activities, like most organizations treat it.
>
> **—H. James Harrington**

Almost every organization should take a deeper look at the states of alignment among their business model, strategy, processes, structures, people, rewards, and implementation. In some cases the findings might prove fairly sobering.

Changes to the strategic parts of the Organizational Master Plan will often drive the need to realign the key parts of the organization so that the Master Plan can be effectively implemented. Even when the individual Annual Operating Plan does not require a major organizational alignment activity, over a period of 5 to 7 years the small individual changes usually add up to the point that a major organizational alignment program has become over due.

Other changes that often justify a major organizational alignment are

- Major process improvement projects
- Major expansions or cutbacks
- Acquisitions
- Entering a new market

There are a number of tools that are used to bring about organizational alignment. Some of the key ones are as follows:

- Business process improvement (BPI)
- Flowcharting
- Knowledge mapping

- Area activity analysis (AAA)
- Organizational change management (OCM)

POLICY DEPLOYMENT (HOSHIN KANRI)

PD is another approach to aligning the organization's goals and strategies that is less thorough than organizational alignment methodology. The two approaches have a lot of activities in common. Policy deployment is a top-down and bottom-up method to ensure that plans and strategies are successfully implemented within the organization and throughout the value chain. It is a planning and alignment method that ties improvement activities, usually requiring breakthrough results, to the long-term strategies of the organization.

Critical success factors are the key things that the organization must do exceptionally well to overcome today's problems and the roadblocks to meeting the vision statements.

Overall, PD requires five basic components insofar as organizational alignment is concerned:*

- **Understanding the customer:** This component is the alignment focus of Strategic Planning—analyzing customer needs, competitor's position, and environmental forces.
- **Goal setting:** This component involves reviewing past performances and creating new directions on objectives, critical success factors, and targets, while changing work processes to close the gaps between targets and desired performance.

* This premise and accompanying analysis is very clearly brought together by Richard Tabor Greene. 1993. New York: in *Global Quality*, pp. 832–839. The major premise is alignment and unification; that is, to use the five basic tasks of PD and Quality Function Deployment (QFD) to eliminate dissipation of those methods into complex procedures and committee work. This concept was developed in the Japanese manufacturing industry during the 1980s and Ford pioneered it in this country about ten years ago. These tasks can be used as one process going in two directions. The ultimate goal is to unify and align the four deployments of QFD, policy, automation, and technology as one process aligned in four directions.

- **Catch-ball within the organization:** This component involves the back-and-forth, up-and-down deployment of new goals and objectives between operating areas and upper management, and negotiating (playing catch-ball) with current process capabilities to allow for new levels of achievement. Alignment catch-ball is a process where upper-level mission and goals are handed down to a lower-level unit; the lower-level unit then evaluates which parts of the mission and goals apply to their unit, and develops their own supporting mission statements with supporting goals.
- **Monthly and quarterly management activities:** This component includes activities that are done in order to measure and track how much of the year's objectives are being accomplished.
- **Checking, inspecting, and problem solving:** This component involves the analysis and solving of problems involved in both daily and monthly management.

Policy deployment alignment starts with the top management team, which is responsible for developing and communicating the vision, as well as building an organizationwide commitment to achievement. The vision is then aligned and deployed through the development and execution of annual policy statements. Through the catch-ball activities previously mentioned, all levels of employees actively participate in generating strategies and action plans for the attainment of the vision and associated objectives. Thus, at each level, progressively more and more detailed action plans are developed and aligned to make the policy statements come alive.

Policy deployment starts with the CEO/president in the form of a few critical policy statements. Each organizational level aligns priorities on the areas needing significant improvement or breakthrough. Each policy statement contains an objective statement, a goal or target, a strategy, a measure, and the person responsible for achievement. As plans are aligned and cascade downward, there should be a clear link between the vision and common goals throughout the organization. It is this focused alignment on a few critical goals that achieves the breakthrough desired, by the marshaling of critical resources in a focused, concentrated manner.

The main outcome of policy deployment is to align the organizational forces of the extended enterprise in the same direction so that everyone within the enterprise knows and understands the overall direction of the organizations involved and how each can play its part to best support that direction.* This includes the selection and alignment of suppliers who share the vision of the organization, and are willing to work together to reduce cycle time and prioritize its improvement efforts tied to the vision. The two factors that can influence this prioritization are importance and the need to improve.

The use of critical success factors (CSFs) can greatly assist with the prioritization, alignment, and focusing of the business design model, and should be viewed as the strategic enabling arm of policy deployment. The missing ingredient in many situations is the inability to correctly understand the customer needs and wants, and to listen to the voice of the customer. This must be addressed in Phase I, Strategic Plan of the Organizational Alignment Cycle.

In order to deal with change, companies and autonomous business units need an effective business design that allows them to react rapidly and continuously, innovate ceaselessly, and take on new strategic imperatives faster and more comfortably, a design based upon organizational alignment or policy deployment.

> Only 5% of the organizations in the West truly excel. Their secret is not what they do – it is how they do it.
>
> **—H. James Harrington**

---

* These authors represent divergent yet complementary viewpoints concerning the implementation strategies and ultimate value of policy deployment. In *Hoshin Planning: The Developmental Approach* (GOAL/OPC, 1989), Bob King discusses the origins and Americanization of hoshin planning. Yoji Akao, in his masterwork *Hoshin Kanri-Policy Deployment for Successful TQM* (Productivity Press, 1992), approaches the subject from its developmental origins in the practices of Japanese administrative business management. Finally, both Bruce Sheridan and Brendan Collins (in related writings) offer a practical working model designed for Western management, which is based upon the FPL Policy Deployment approach to organizational alignment. Bruce Sheridan, *Policy Deployment: The TQM Approach to Long- Range Planning* (Milwaukee, WI: ASQ Quality Press, 1993) and Brendan Collins, *Management by Policy: How companies focus their total quality efforts to achieve competitive advantage* (Milwaukee, WI: ASQ Press, 1993).

2

What's in an Effective Planning Process?

> We make one of two kinds of mistakes: We plan too much and do too little. We plan too little and do the wrong things.
>
> **—H. James Harrington**

The master planning for an organization doesn't really have a definite starting and ending point, particularly in today's ever-changing business environment. As pointed out earlier, the winners really try to gain a significant advantage through the use of an effective planning approach. Some of these plans are very long, elaborate affairs, and some are quicker-hitting events. By focusing on being a leader in their industry, they are creating the changes as opposed to always reacting to change. In this regard, defining an appropriate planning process is critical. Today, the planning processes need to be adjusted to consider many factors, such as the current position of the organization (are they the leader or are they in trouble), the style of the management team, and time and scheduling. However, the most successful planning processes have many attributes in common. This even holds true independent of the form the plan takes, although they are essential to the success of developing the plan content discussed earlier. These attributes can be summarized as follows:

- In-depth consideration of customers, competitors, and capabilities
- Involvement of implementers as planners
- Identification of obstacles to success and related critical success factors
- Testing against resource availability (time, money, and people)
- Development of specific communications plans
- Consensus on how the plan will be used in the implementation efforts

> If I had eight hours to chop down a tree, I'd spend six sharpening my axe.
>
> **—President Abraham Lincoln**

IMPLEMENTERS AS PLANNERS

The winners know that even the best plan, not implemented, is useless. They work very hard to develop a useful planning process, which requires that the management team responsible for its execution is involved in the development of the plan. This not only taps into their experience but it improves both the level of buy-in and the understanding of the organization's direction and initiatives. Although consensus is desired, the winners know that unanimity is seldom possible, but understanding and support is. In contrast, the losers don't involve line management and prepare the plan in a vacuum. They run the risk of both making poor strategic decisions and having poor execution. No wonder they're the losers!

> All the flowers of all the tomorrows are in the seeds of today.
>
> **—Indian Proverb**

We liken the planning process to farming (see Figure 2.1) as follows:

- The Strategic Plan is like plowing the field.
- The Improvement Plan is like planting the seeds.
- Funding and manning the Master Plan is like the fertilizer and rain that nurture the plants.
- Measuring the results is like harvesting the food.
- Analyzing the results is a lot like saving seeds for next year's planting.

> When planning for a year, plant corn. When planning for a decade, plant trees. When planning for life, train and educate people.
>
> **—Chinese Proverb**

A good Organizational Master Plan has the following elements:

- A clear focus on specific markets and customers to be served
- A vision of what the future will be like in terms of the market and the organization's leadership role in it
- A mission for the organization
- A clear articulation of the most important strategic initiatives or focus that the organization must undertake

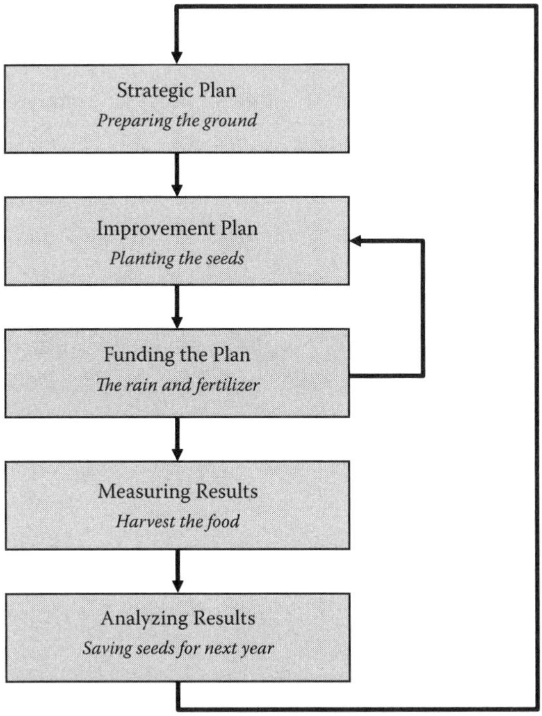

FIGURE 2.1
Growing a successful Organizational Master Plan.

- Simple measurements or objectives related to the strategic initiatives
- Well-defined strategies for accomplishing these elements and identification of the responsible leaders.
- Organizational values and a culture that will support and enhance the implementation of the chosen strategy

In fact, most plans can probably be a lot shorter and more precise than they are today. Far too many of us have had the experience of seeing large, voluminous plans that are prepared regularly but are never used. These "dust collectors" often do a reasonable job of describing everything, but do a terrible job of providing focus for the organization. A clear, simple set of strategic decisions for competing in a complex world will result in a core focus that can be easily summarized and communicated throughout the organization. Winners follow the planning rule, *decisions, not descriptions*, while the losers tend to follow *descriptions, not decisions.*

> Have a plan. Follow the plan, and you'll be surprised how successful you can be. Most people don't have a plan. That's why it's easy to beat most folks.
>
> **—Paul Bear Bryant, former University of Alabama Football Coach**

THE CUSTOMER IS KING

All types of business plans have only one purpose—to satisfy the customer's requirements. These requirements may be documented, verbal, or even unknown by the customer himself or herself. Future (strategic) plans need to consider what the customer will want two, five, or even ten years from now. This means that you will need to get into your time machine and advance your thinking by ten years. One of the biggest mistakes you can make is to believe that the customer knows your technology well enough to predict what types of services/products it will be creating in the future. We like to classify customer requirements in the following five ways:

- Articulated requirements—what they tell you they require
- Observed requirements—what you see that they require
- Hidden requirements—needs that the customer has but has a hard time defining
- Latent requirements—requirements that are not needed now but will be needed in the future
- Technology-driven requirements—needs that new technology developments will trigger

The problem, which many planners encounter, is that they only recognize the articulated requirements, and as a result, they are unable to predict future customer requirements. These organizations live in a reactive mode, always trying to catch up, always running behind the leaders in the industry, picking up the leftovers from the really successful organizations.

HOW TO GET CUSTOMER REQUIREMENTS

There are six main ways of getting customer requirements. Each of these has advantages and disadvantages. They are

1. Focus groups
2. In-depth interviews
3. Observational interviews
4. Projective testing
5. Leading-edge groups
6. And of course, the most basic—surveys

Focus Groups

Focus groups are small gatherings of 8–10 people who discuss a specific set of topics to define the situation.

- Advantages:
  - Cost effective
  - Interactive
  - Productive
  - Open atmosphere
- Disadvantages:
  - May not be factual
  - Can be biased
  - Can get misdirected
  - Strong individuals can lead the group to their point of view

In-Depth Interviews (IDIs)

These are one-on-one interviews using a prepared set of questions designed to obtain in-depth insight into the customer's culture and requirements.

- Advantages:
  - No group pressure
  - Can be time-consuming
  - Confidential
  - Open discussion
  - In-depth discussion
- Disadvantages:
  - Costly
  - Time-consuming
  - Lack of group dynamics

Observational Interviews

These are observations of the customer's characteristics and activities when they are using the product or service.

- Advantages:
  - You see what they do
  - Little bias
  - Hidden requirements are identified
- Disadvantages:
  - Costly
  - Time-consuming
  - Depends on the individual customer

Projective Testing

This is used to define hidden and latent requirements. It uses approaches like word association and fill-in-the-blank exercises.

- Advantages:
  - One-on-one discussions
  - In-depth insight into customer feelings
  - You can discuss something or someone else
- Disadvantages:
  - Very subjective
  - Expensive
  - Time-consuming

Leading-Edge Groups

These are workshops or focus groups with customers who are ahead of the average or mainstream customers. These are technology discussions that are designed to define how the future technology will drive future customer needs (typically 2–3 days per session).

- Advantages:
  - Identify emerging requirements
  - Define breakthrough ideas and concepts

- Disadvantages:
  - Hard to set up
  - Very costly
  - Time-consuming
  - Difficult to interpret the results

Surveys

Surveys can be conducted by direct interface with the individual or by indirect contact (for example, mail, Internet, etc.). The survey design plays a critical part in the accuracy of the results. The sample selection can also cause the results to be inaccurate. Both of these factors need to be considered when designing and conducting the survey.

- Advantages:
  - Cost-effective
  - No group pressure
  - Confidential
  - Large sample size
  - Productive
- Disadvantages:
  - Depends on individual customers
  - Can get misdirected
  - Important things may not be identified

UNDERSTANDING YOUR CUSTOMER'S INTERFACE

The following ten questions will help you understand your customer and potential customers better than you have ever understood them before.

1. Why do customers need your product or service?
2. How do they find you or your competitor's products and services?
3. What do the customers use to select the products or services they buy?
4. How do they buy the products or services?
5. What is the delivery process?

6. How is the product put to use?
7. How is the product or service actually used on a day-to-day basis?
8. What problems are customers having in using your or your competitor's products or services?
9. What are the problems with repairing or maintaining the products?
10. How does the customer dispose of the product?

EXAMPLE OF CUSTOMER-FOCUSED CEO

Jacques Nasser, the CEO of Ford Motor Company, is very customer focused and he wants all managers and employees to also be customer focused. As a result, teams of Ford employees went out to customer-insight centers for a course on how to listen and how to talk to customers. They then went out on an eight-week assignment to talk to Ford customers; they discussed subjects like design, safety, comfort, ease of use, even down to types of coffee holders. This is one of the functions that made Ford (by 1999) the world's most profitable automotive maker and produced a 130% increase in stock prices over a two-year period. It is one of the reasons that Ford was the only one of the Big Three automakers in the United States that didn't need bailout money from the government.

MARKET FOCUS

Clear market focus provides an organization and its people with an opportunity to accomplish great things through understanding the customer's needs better than its competitors. This understanding then provides the foundation for the development of superior products and services. Knowing who you're serving is the first step in developing a good Organizational Master Plan and achieving the ultimate business results that the winners in business experience. The losers more often are so unfocused that they cannot serve anyone well with the demands on their limited resources. The really successful organizations will understand markets in terms of what actual customers want and will pay for, thus reaching a level of understanding few others achieve. As one leading computer manufacturer said, You certainly couldn't argue with their results!

ORGANIZATIONAL MASTER PLAN DRIVERS

Various business analysis techniques can be used in developing an Organizational Master Plan, including SWOT (strengths, weaknesses, opportunities, and threats) and PEST (political, economic, social, and technological analysis) or STEER involving sociocultural, technological, economic, ecological, and regulatory factors.*

In the 1960s, the words *long-range planning* were used to describe the strategic planning system. Other names have subsequently been coined, some of the different terms are: comprehensive corporate planning, comprehensive managerial planning, total overall planning, long-range planning, formal planning, comprehensive integrated business planning, corporate planning, strategic planning, and other combinations of these words and terms. More and more, however, formal strategic planning is used to describe what is usually meant when these phrases are employed.

Most writers in the field have their own pet definitions of the terms listed in the preceding paragraph. They vary greatly in terms of level of abstraction, substance, and general acceptance. Most writers agree, however, that strategic planning should be described from several points of view. For a deeper understanding, we define the Organizational Master Plan from four points of view or lenses, each of which is needed to understanding it.†

- **Point of view 1:** Planning deals with the *future focus* of current decisions. This means that strategic planning looks at the chain of cause-and-effect consequences over time of an actual or intended decision that a manager is going to make. If the manager does not like what is seen ahead, the decision can and should be changed. The strategic planning process also looks at the alternative courses of action that are open in the future, and when choices are made among the alternatives, they become the basis for making current decisions. The essence of formal strategic planning is the systematic identification of opportunities and threats that find their home in the future. This is in combination with other relevant data, which provides a basis for an organization making better current decisions to exploit

* From *Wikipedia*, the free online encyclopedia.

† Based on the works of George Steiner and other planning gurus, as described in Steiner's master work *Strategic Planning: What Every Manager Must Know* (New Tork: The Free Press/Macmillan Publishing, 1979).

the opportunities and to avoid threats. Strategic improvement planning means designing a desired future and identifying ways to bring it about.

- **Point of view 2:** The Organizational Master Plan is a systematic process. The process starts with the setting of organizational imperatives, defines strategies and policies to achieve them, and develops detailed plans to make sure that the strategies are implemented in order to achieve the desired future state. Next, it is a process of fast-forward deciding what kind of planning effort is to be undertaken, when it is to be done, how it is to be done, who is going to do it, and what will be done with the results. Strategic planning for most organizations results in a series of plans generated over a specified period of time, set aside for the alignment of the plans.* However, it should also be conceived as a continuous process, especially with respect to strategy formulation, because changes in the business environment are continuous, requiring the Organizational Master Plan to be adjusted. The idea here is not that plans should be changed every day but that thought about planning must be continuous and supported by appropriate action when necessary. The Organizational Master Plan should have enough flexibility to be adjusted to cover the week-to-week changes within the guidelines set forth in the Strategic Plan.
- **Point of view 3:** The Organizational Master Plan process deals with an attitude, a philosophy, a way of life. Planning necessitates dedication to acting on the basis of considering the future possibilities, and a determination to plan constantly and systematically align it as an integral part of management. Strategic planning is more of an alignment thought process, and an intellectual exercise, than a prescribed set of processes, procedures, structures, or techniques. "For best results, managers and staff in an organization must believe Strategic Planning is worth doing and must want to do it as well as they can.

* In the words of Aaron Wildaysky: "Planning may be seen as the ability to control the future consequences of present actions. The more consequences one controls, the more one has succeeded in planning. Planning is a form of causality. Its purpose is to make the future different from what it would have been without this intervention. Planning therefore necessitates a causal theory connecting the planned actions with the desired future results. Planning also requires the ability to act on this theory; it requires power. To change the future, one must be able to get people to act differently than they otherwise would. The requirements of successful planning from causal theory to political power, grow more onerous as its scope increases and the demands for simultaneous action multiply at a geometric rate." See Aaron Wildaysky, "Does Planning Work?" *Public Interest*, summer 1971, p. 101.

Not to do it well is not a sin," says Ackoff, "but to settle for doing it less than well is."*

- **Point of view 4:** The Organizational Master Plan cycle links three major types of plans together. They are the Strategic Business Plan, Strategic Improvement Plan, and Annual Operating Plan. In an enterprise with subsidiaries and decentralized divisions, there may be this type of alignment in each division's plans and a different linkage between the Organizational Master Plan made at headquarters and divisional plans. It is through the alignment linkages that senior management's strategies are translated into current decisions.

> Organizing is what you do before you do something, so that when you do it, it is not all mixed up.
>
> **—A. A. Milne**

The concept of a hierarchical structure of plans is also contained in this definition: "Strategic Planning is the systematic, formalized effort of a company to establish basic company purposes, objectives, policies, and strategies and to develop detailed plans to implement policies and strategies to achieve objectives and basic company purposes."†

> Every employee should have and know what his/her role is in implementing the Organizational Master Plan.
>
> **—H. James Harrington**

* Russell L. Ackoff wrote, "[P]lanning is the design of a desired future and of effective ways of bringing it about." See *A Concept of Corporate Planning* (New York: Wiley Interscience, 1969).

† Russell L. Ackoff interview by Frank Voehl, Guardian Life Insurance Company, NYC, April, 1978.

3

The Organizational Master Plan

> The nice thing about not having an Organizational Master Plan is that you can't be off course if you do not know where you are going.
>
> **—H. James Harrington**

INTRODUCTION TO THE ORGANIZATIONAL MASTER PLAN

The *Organizational Master Plan* is the combination and alignment of an organization's Business Plan, Strategic Business Plan, Strategic Improvement Plan, Strategic Plan, and Annual Operating Plan.

Let's start out by answering two questions. Rate these using a scale of 1 to 10 with one being extremely low and ten being extremely high.

1. How important is strategic planning to the organization?
2. Based on actual performance, how effective have the strategic planning activities been to date?

> Good fortune is what happens when opportunity meets with planning.
>
> **—Thomas Edison**

For most people there is a major difference between the answers to questions 1 and 2. Ask yourself why there is difference? Take the time to make a list of why you think there is a difference.

The Organizational Master Plan combines the five major management plans together into a homogeneous, agreed-to, focused approach to managing the organization's culture, business, and operations. These major management plans are

- The Business Plan (BP)
- The Strategic Business Plan (SBP)
- The Strategic Improvement Plan (SIP)
- The Strategic Plan (SP), which is the prioritized integration of the Strategic Business Plan (SBP) and the Strategic Improvement Plan (SIP).
- Annual Operating Plan (AOP)

- **Business Plan (BP):** A Business Plan is a formal statement of a set of business goals, the reasons why they are believed to be attainable, and the plan for reaching those goals. It also contains background information about the organization or team attempting to reach those goals.
- **Strategic Business Plan (SBP):** The strategic business plan focuses on what the organization is going to do to grow its market. It is designed to answer these questions: What do we do? How can we beat or avoid competition? It is directed at the product and/or services that the organization provides as viewed by the outside world. When an organization is just being funded, this plan is often just called a Business Plan.
- **Strategic Improvement Plan (SIP):** The Strategic Improvement Plan focuses on how to change the culture of the organization. It is designed to answer these questions: How do we excel? How can we increase value to all the stakeholders? It addresses how the controllable factors within the organization can be changed to improve the organization's reputation and performance.
- **Strategic Plan (SP):** The Strategic Plan is a document that is the result of strategic planning. It defines the organization's strategy and/or direction and makes decisions on the allocation of resources in pursuit of the organization's strategy including its capital and people. It focuses on the future of the organization and is the combination of the Strategic Business Plan and the Strategic Improvement Plan with each item prioritized to maximize the organization's performance. The Strategic Plan is the combination and prioritization of the individual strategies of the Strategic Business Plan and the Strategic Improvement Plan.

- **Annual Operating Plan (AOP):** The Annual Operating Plan is a formal statement of business short-range goals, the reasons they are believed to be attainable, the plans for reaching these goals, and the funding approved for each part of the organization (budget). It includes the implementation plan for the coming years (years 1–3) of the Strategic Plan. It may also contain background information about the organization or teams attempting to reach these goals. One of the end results is performance plans for each manager and employee who will be implementing the plan over the coming year. The Annual Operating Plan is often just referred to as the Operating Plan (OP).
- **Strategic Management:** Strategic management is the process of specifying the organization's mission, vision, and objectives, developing policies and plans, often in terms of projects and programs that are designed to achieve these objectives, and then allocating resources to implementation of policies and plans, projects, and programs.

The Organizational Master Plan must combine all of the major planning processes together and, at the same time, ensure they are customer focused. All of the plans must be based upon understanding the customer requirements and preferences, plus having an excellent understanding of what the competition is doing currently and will be doing in the future. Figure 3.1 is a model of an Organizational Master Plan, indicating the five different parts of this plan: the Business Plan, the Strategic Business Plan, the Strategic Improvement Plan, the Strategic Plan, and the Annual Operating Plan. (The Strategic Plan is the prioritized combination of the Strategic Business Plan and the Strategic Improvement Plan; in the model it is represented by the combination of the two gray diamonds.)

The Business Plan's primary objective is to inform external investors. It is prepared early in the organization's development cycle and it is often used to provide potential investors with information related to the organization and its management. Information from it feeds directly into the Strategic Plan. While the Strategic Plan focuses on the future of the organization, the focus of the Annual Operating Plan is short range and addresses the current operations. You will also note that the Strategic Business Plan and the Annual Operating Plan define the direction that the organization wants to take and complement each other. On the other hand, the Strategic Improvement Plan focuses on the organization's operations (how things get done) and its culture.

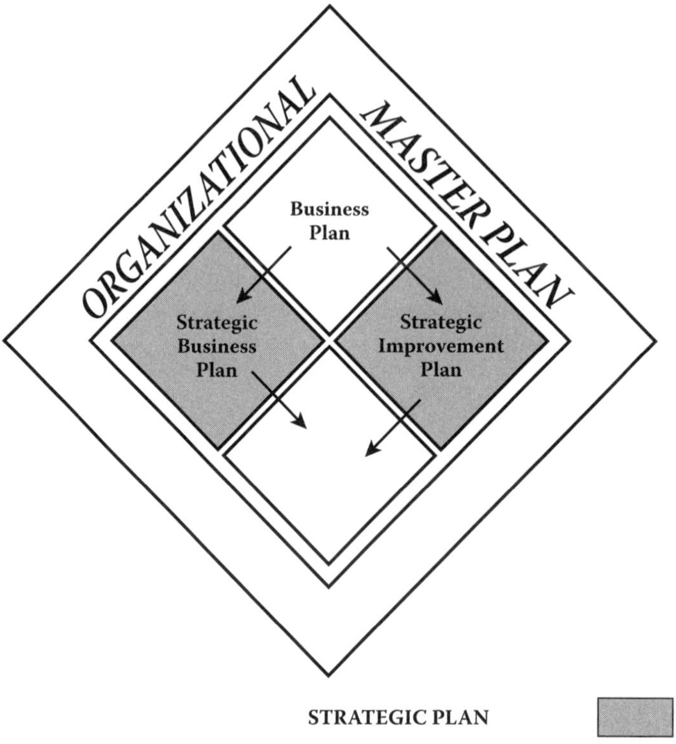

FIGURE 3.1
Five parts of the Organizational Master Plan.

One major objective of this book is to help you close the gaps among the five different plans.

> Good plans shape good decisions. That's why good planning helps to make elusive dreams come true.
>
> **—Lester R. Bittel, *The Nine Master Keys of Management***

INTRODUCTION TO STRATEGIC PLANNING

Over the past several years there has been a lot of leading-edge thinking and many definitions put forth of how various areas of organizational strategy approaches might be viewed in the short and long term, and how they might be classified. Regardless of the definitions applied, what is common among all the successfully aligned organizations is their ability to identify those four or five key areas of short-term strategic focus that are characterized by the following:

- Customers value the benefits that the focus provides.
- Concentration of resources toward being the absolute best in your chosen areas of emphasis will enable you to excel.
- Excellence in these areas will be difficult for competitors to imitate.
- These areas of focus are your organization's capabilities or what you're really good at, not outcome measures like market share, profit margin, and so on.

In an August 26, 1996 article, *Businessweek* magazine reported the following:

> A recent study by the Association of Management Consulting Firms found that executives, consultants, and B-school professors all agree that organizational strategy is now the single most important management issue and will remain so for the next five years. "We are seeing strategy make a rebound," says Vijay Govindarajan, a well-known strategy professor at Dartmouth College's Amos Tuck School of Business Administration. "Strategy has become a part of the main agenda at lots of organizations today." But if Strategic Planning is back with a vengeance, it's also back with a difference.
>
> Gone are the abstraction, sterility, and top-down arrogance of the old model. The death knell for that approach was sounded in 1983 when General Electric Chairman John F. Welch dismantled the company's once heralded planning department, where as many as 200 senior-level staffers used to crank out vinyl-bound reports. Welch found GE planners too consumed with operating and financial details instead of competitive positioning and the creation of future markets and too divorced from the day-to-day reality of line managers.
>
> Today's gurus of strategy urge organizations to democratize the process—once the sole province of an organization's most senior officers—by handing Strategic Planning over to teams of line and staff managers from different disciplines. Frequently, these teams include junior staffers, hand picked for their ability to think creatively, and near-retirement old-timers willing to tell it like it is. And to keep the planning process close to the realities of markets, today's strategists say it should also include interaction with key customers and suppliers.

> Planning … is anticipatory decision making. It is a process of deciding … before action is taken.
>
> **—Professor Russell Ackoff, Wharton Business School**

THE ORGANIZATIONAL MASTER PLAN OVERVIEW

- 90% of organizations feel it is important to implement their Strategic Plan
- 85% of the senior managers spend less than one hour per month on the Strategic Plan
- 60% don't link the Strategic Plan to the budget
- 25% only have rewards linked to performance
- 5% of staff understand the Strategic Plan*

To be successful, all organizations need to have an effective approach that sets the course for today, as well as future decisions, and to prioritize its activities. It should be made up of three parts (see Figure 3.2).

1. Setting directions
2. Defining expectations
3. Defining actions that need to be taken

These purposes are interrelated and react with each other, and each drives a number of outputs that communicate its intent to the stakeholders. Figure 3.3 shows the three major purpose areas that each of the types of plans is designed to impact.

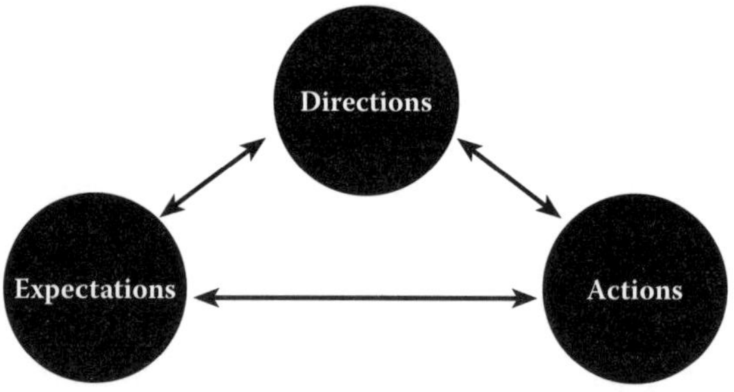

FIGURE 3.2
The three purposes of an Organizational Master Plan.

* *Resource Management Excellence: The Art of Excelling in Resource Management* (Paton press, 2007) 131, 150.

	Direction	Expectations	Actions
Business Plan	X	X	X
Strategic Plan	X	X	
Strategic Business Plan	X	X	
Strategic Improvement Plan	X	X	
Annual Operating Plan		X	X

FIGURE 3.3
The different types of plans and their major impact on the three purposes of the planning cycle.

Purpose	Outputs	Frequency of Updates
Setting Direction	☐ Mission	☐ Seldom
	☐ Long-Range Vision Statement	☐ 10–25 years
	☐ Short-Range Vision Statement	☐ 5–10 years
	☐ KBD's Vision Statement	☐ 5 years
	☐ Value Statements	☐ 5 years
	☐ Strategic Focus	☐ 3 years
	☐ Critical Success Factors	☐ 3 years
Defining Expectations	☐ Organizational Objectives	☐ 5–10 years
(Measurements)	☐ Performance Goals	☐ 1–5 years
Defining Actions	☐ Strategies	☐ 1–5 years
	☐ Tactics	☐ 1–3 years
	☐ Budgets	☐ 1–3 years
	☐ Personnel Performance Plans	☐ 3–12 months

FIGURE 3.4
Organizational Master Plan elements and frequency of changing them.

An Organizational Master Plan is a communication, planning, and business strategies that reaches and involves every employee from the board to the boiler room in support of common goals and objectives. It is a three-way interactive process that provides direction, expectations, and funding. It also defines activities required to meet the agreed-upon expectations. An Organizational Master Plan includes the twelve outputs (see Figure 3.4). To have a good Master Plan, the organization must understand and do an excellent job of creating and communicating these twelve planning outputs.

SETTING DIRECTION

By failing to prepare, you are preparing to fail.

—Benjamin Franklin

The principle role of top management is to set the direction for the organization. This can best be accomplished and communicated to the stakeholders through the Organizational Master Plan. The outputs that are used to provide this direction are

- **Mission:** The mission is the stated reason for the existence of the organization. It is usually prepared by the chief executive officer and seldom changes, normally only when the organization decides to pursue a completely new market.

  Example: An example of a "to be" mission from Boeing is: *Our long-range mission is to be the number one aerospace company in the world, and among the premier industrial firms, as measured by quality, profitability and growth.*

  Example: An example of a "to do" mission from McDonald's is: *To satisfy the world's appetite for good food, well-served, at a price people can afford.*

- **Vision:** The vision is a documented or mental description or picture of a desired future state of an organization, process, team, key business drivers, or activity.
- **Key business drivers (KBD) (also called *controllable factors*):** KBDs are the things within the organization that management can change that control and/or influence the organization's culture and/or the way the organization operates.

  Typical KBDs that impact organizational performance are listed here. Those with an asterisk (*) have pre-defined twelve-level maturity grids already developed for them.

  1. Measurement system*
  2. Training*
  3. Management and leadership methods*
  4. External customer partnership interface*
  5. Supplier partnership*
  6. Business processes*

7. Production processes
8. Corporate interface to the divisions
9. Employee partnership
10. Research and development activities
11. After-sales service processes
12. Knowledge Management*

- **Key business driver (KBD) vision statements:** KBD vision statements (also called *controllable factor vision statements*) are statements of how the key business drivers will be operating five years in the future. They are developed during the strategic improvement planning process. There are usually 8 to 12 of these vision statements.

  Example of a processes vision statement: *Our critical processes provide a competitive advantage that employees and customers recognize as world class. Continuous improvement using statistical data is a way of life. Processes are defined with assigned owners, evaluated against established measures, and operate in the real world.*

- **Vision statements:** These are prepared by top management and are directed at what the organization's output will be like and/or how it will be used 10–25 years in the future. Some organizations have a long-term vision statement (10–25 years) and a short-term vision statement (5–10 years).

  Ford's vision statement: *To become the world's leading consumer company for automotive products and/or services.*

- **Values:** These are things that an organization holds as sacred; things that should never be sacrificed. Much care should be taken when an organization selects its values to be sure they do not conflict with each other. Values are often interchanged with beliefs, but management is held accountable for complying with values.
- **Value statements:** Value statements are documented directives that set behavioral patterns for all employees. These are the basic beliefs that the organization is founded upon, the principles that make up the organization's culture. They are deeply engrained operating rules or guiding principles of an organization that should not be compromised. They are rarely changed. Value statements are sometimes called *operating principles, guiding principles, basic belief operating rules.*

Owens Corning Fiberglas uses *guiding principles* in place of values. Their guiding principles are

- Customers are the focus of everything we do.
- People are the source of our competitive strength.
- Involvement and teamwork is our method of operation.
- Continuous improvement is essential to our success.
- Open, two-way communication is essential to the improvement process and our mission.
- Suppliers are team members.
- Profitability is the ultimate measure of our efficiency in serving our customer's needs.

- **Strategic focus:** These are the key things that will set the organization apart from its competitors over the next five years. This list is defined by top- and middle-level managers.
- **Critical success factors:** These are the key things that the organization must do exceptionally well to overcome today's problems and the roadblocks to meeting the vision statements.
- **Core competencies:** These are the technologies and production skills that underlie an organization's products or services (e.g., Sony's skill at miniaturization).
- **Core capabilities:** These are the business processes that visibly provide value to the customer (e.g., Honda's dealer management processes).
- **Strategic excellence positions:** Unique and distinctive capabilities that are valued by the customer and provide a basis for competitive advantage (e.g., Avon's distribution system).

DEFINING EXPECTATIONS (MEASUREMENTS)

One of the major purposes of an Organizational Master Plan is to define what management and the stockholders expect from the organization's performance over the next 5–10 years and then to communicate how success will be measured. The outputs that are used to communicate these expectations are

- **Organizational objectives:** Organizational objectives are used to define what the organization wishes to accomplish over the next 5–10 years. These are usually defined by top management.

Example: IBM released the following objectives that it planned to accomplish during a 10-year period.

- To grow with the industry.
- To exhibit product leadership across our entire product line. To excel in technology, value, and quality.
- To be the most effective in everything we do. To be the low-cost producer, the low-cost seller, the low-cost administrator.
- To sustain our profitability, which funds our growth.

- **Performance goals:** Performance goals are used to quantify the results that will be obtained if the organizational objectives are satisfactorily met. They include a time frame for the goal and the magnitude of the targeted change.

  Example: On February 25, 1986, then president of the United States Ronald Reagan released Executive Order 12552, stating: "There is hereby established a government-wide program to improve the quality, timeliness, and efficiency of services provided by the federal government. The goal of the program shall be to improve the quality and timeliness of service to the public, and to achieve a 20% productivity increase in appropriate functions by 1992. Each executive department and agency will be responsible for contributing to the achievement of this goal."

How often have you heard the saying, "Anything worth having is worth working for"? So it goes with business or strategic planning, but how can we tell if our efforts are paying off? We can surely see sales, profits, and cash flow, all tangible measures of current business results. Both winners and losers utilize these very traditional measures. However, the winners tend to also track some very simple measures or objectives related to their progress in developing long-term competitive strength in their chosen areas of strategic focus. For example, if unsurpassed service is one of the areas of strategic focus, the management of the winners would have regular reporting on a few simple measures of unsurpassed service. These measures might even include things like the number of customer referrals, in addition to measures like customer problem resolution cycle times and the customer retention rate. This linking of very visible, regularly communicated measures to the strategic focus of the organization is crucial to the transition process required to make that vision become a reality. Losers tend to only rely on traditional financial measures to gauge their success.

Another reason why selecting and defining measures is important is their value in informing the organization of the pace of change and implementation required. Organizations that expect quantum leaps in performance against measures will need to prioritize and focus their resources in support of these stretch targets. Winners exhibit a keen flexibility and an intuitive sense of how far to stretch, and seldom set the easy-to-beat or absolutely unrealistic targets that losers tend to use. As we saw with Hewlett-Packard (HP) when John Young set aggressive targets (e.g., 10 times improvement in a critical business process within 10 years), HP used this to substantially reprioritize the way they used their resources and thereby created a compelling challenge for their organization to rally around.

> *Example A:* Motorola set "stretch" objectives for itself. William J. Weisz, chief operating officer of Motorola, explained: "In 1981, we developed as one of the top ten goals of the company, the 5-Year, Tenfold Improvement Program. This means that no matter what operation you're in, no matter what your present level of quality performance, whether you are a service organization or a manufacturing area, it is our goal to have you improve that level by an order of magnitude in 5 years."*

DEFINING ACTIONS

Another purpose of an Organizational Master Plan that drives the organization's change process is to define the actions that will be taken to implement the plan over the next five years. It is designed to focus the resources of the organization in line with its expectations. The outputs that are used to communicate these actions are

- **Strategies:** Strategies define the approaches that will be used to meet the performance goals.
- **Tactics:** Tactics define how the strategies will be implemented. They explain how the strategies will be accomplished.
- **Budgets:** Budgets provide the resources required to implement the tactics.
- **Performance plans:** Personnel performance plans are contracts between management and the employees that define the employees' roles in accomplishing the tactics, and the budget limitations that the employees have placed upon them.

* H. James Harrington, *The Quality/Profit Connection* (Strategic Planning, 1989), 89.

Purpose	12 Outputs	Strategic Plan	Annual Operating Plan
Setting Direction	Mission	X	
	Short-term visions	X	X
	Long-term visions	X	
	Values	X	
	Strategic focus	X	X
	Critical success factors	X	X
Defining Expectations (Measurements)	Organizational objectives	X	X
	Performance goals	X	X
Defining Actions	Strategies	X	X
	Tactics		X
	Budgets		X
	Performance Plans		X

FIGURE 3.5
Two parts of an Organizational Master Plan.

Figure 3.5 shows how the two plans relate to the twelve planning outputs, which make up the Organizational Master Plan.

SUMMARY

It is easy to see that the Organizational Master Plan is designed to bring all the major plans together into one common plan that unites all the parts of the organization. It keeps operating approaches and strategic vision marching hand in hand.

> The Organizational Master Plan brings the top hats and hard hats together.
>
> **—H. James Harrington**

4

Change Management as Part of the Organizational Master Plan

Change management approaches may sound like common sense but, too often, common sense is not commonly practiced.

—H. James Harrington

Only the very simplest parts of the Organizational Master Plan project will be successful without being accompanied by a very effective organizational change management process. A typical Organizational Master Plan project will impact your processes, procedures, staff, organizational structure, and the culture of the organization, putting a great deal of stress on your people. This impact will generate a lot of resistance to the project.

To offset the many problems that occur if the affected employees are not made part of the project before it is implemented, a seven-phase change management methodology has been developed that starts as soon as the project team is assigned (see Figure 4.1).

Unless commitment is made, there are only promises and hopes; but no plans.

—Peter F. Drucker, Writer

The following gives more details related to each of the seven phases.

- **Phase I—Clarify the project:** In Phase I, the scope of the project and the level of commitment by management and the affected employees required for the project to succeed is defined.
- **Phase II—Announce the project:** In Phase II, a tailored Change Management Plan is developed and communicated to all the affected constituents. Preplanning and sensitivity to the unique needs of

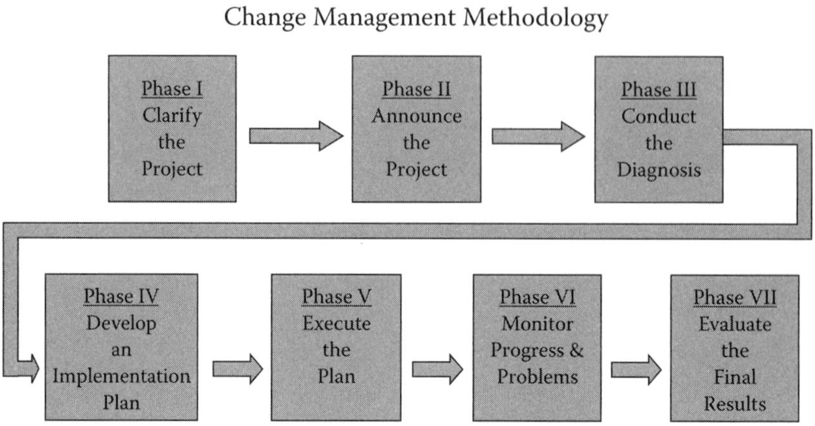

FIGURE 4.1
Seven phases of the change management methodology.

various groups will minimize disruption and set the stage for acceptance of the need for the change.

- **Phase III—Conduct the diagnosis:** During Phase III, surveys and other types of analysis tools are used (for example, a Landscape Survey) to determine what implementation barriers exist that could jeopardize the success of the change. This diagnostic data, coupled with the rich dialogue that occurs during Phase II, provide the basis for developing an effective implementation plan.
- **Phase IV—Develop an Implementation Plan:** The implementation plan defines the activities required to successfully implement the project on time, within budget, and at an acceptable quality level. Typical things that will be addressed in this plan are
  - Implications of status quo
  - Implications of desired future state
  - Description of the change
  - Outcome measures
  - Burning platform criteria
  - Comprehensive or selected application of implementation architecture
  - Disruption to the organization
  - Barriers to implementation
  - Primary sponsors, change agents, change targets, and advocates
  - Tailoring of announcement for each constituency
  - Approach to pain management strategies

    - Actions to disconfirm status quo
    - Alignment of rhetoric and consequence management structure
    - Management of transition state
    - Level of commitment needed from which people
    - Alignment of project and culture
    - Strategies to improve synergy
    - Training for key people
    - Tactical action steps
    - Major activities
    - Sequence of events
- **Phase V—Execute the plan:** The goal of Phase V is to fully achieve the human and technical objectives of the change project on time and within schedule. It is designed to achieve these objectives by reducing resistance and increasing commitment to the project.
- **Phase VI—Monitor progress and problems:** The goal of Phase VI is to keep project implementation on track by consistently monitoring results against the plan.
- **Phase VII—Evaluate the final results:** The intent of Phase VII is to provide a systematic and objective collection of data to determine if the tangible and intangible objectives of the project have been achieved and to provide insight into lessons learned and potential problem areas that may arise in future change projects.

People who welcome change make progress. People who fight it make excuses.

—H. James Harrington

During Phase VII, management should show its appreciation to the natural work teams (NWT) and individuals who expended exceptional effort during the project or who implemented major improvements. The people who are implementing the change need to wear two hats—one is to cover the technical activities required for the project and the other is as a change agent.

> More than 90% of effectively formulated strategies don't get successfully implemented.
>
> **—Advait Kurlekar Director, Cedar Consulting**

We agree with Advait Kurlekar, and our data indicates that the primary reason for these failures is not preparing the people who are affected by the change so that they are willing to change their behavioral pattern in keeping with the modifications required by the change (organizational change management).

Note: For more information on change management, we recommend reading *Project Change Management* by H. James Harrington, Daryl R. Conner, and Nicholas L. Horney (New York: McGraw-Hill, 2000) and *Change Management Excellence* by H. James Harrington (Chico, CA: Paton Press, 2006).

CAN THE ORGANIZATION HANDLE THE REQUIRED CHANGE?

If organizational change could be limited to only machines or physical facilities, implementation would be relatively simple. Most major changes within the human work environment, however, require employees to modify something about the way they think, feel, or behave. The addition of these human factors significantly increases the complexity of the change process. For this reason, successful implementation of change requires an understanding of the human as well as the technical aspects involved in the situation.

> Men often oppose a thing merely because they had had no agency in planning it, or because it may have been planned by those whom they dislike.
>
> **—Alexander Hamilton, U.S. Founding Father**

IMPLEMENTATION HISTORY ASSESSMENT

One of the key factors that affects an organization's ability to implement present or future change is its past implementation success. Previous experience is a potent predictor of the future. Prior implementation difficulties are likely to be repeated when engaging in new projects. Therefore, an assessment of previous implementation experience is critical in planning for future changes.

To do this evaluation, we suggest doing an Implementation History Assessment. The following are typical statements that would be evaluated as to how consistent the statement is with how the organization has implemented major changes in the past.

- This organization has a poor track record for identifying and resolving problems during change.
- Risk taking has been discouraged and creative ideas have been ignored.
- Too much emphasis has been placed on finding who caused the problem and punishing that person.
- We usually have not allowed sufficient time for carrying out changes.
- Past change activities have been poorly monitored by managers.
- Past change efforts have not been communicated effectively down through the organization, resulting in people feeling confused about how the change affects them or what they should do differently.

The results of the analysis can be used to define the risks related to twenty-five key characteristics that are critical to successful implementation of change within an organization. These results will be used to help define the magnitude of the changes that can be included in the Organizational Master Plan. Mitigation plans should be developed for all the high-risk barriers to successful implementation of the Strategic Plan.

> Each change can be your stepping stone to a better future for you, or your tombstone. It is up to you and how you react to the change.
>
> **—H. James Harrington**

5

Trend Analysis

Tomorrow's problems can be avoided by understanding yesterday's results.

—H. James Harrington

The four points of view presented in Chapter 4 will serve as a basis for the development of the concepts described throughout this chapter. They were

1. Planning deals with the *future focus* of current decisions.
2. The Organizational Master Plan cycle is a systematic process.
3. The Organizational Master Plan process deals with an attitude, a philosophy, a way of life.
4. The Organizational Master Plan cycle links three major types of plans together.

Before proceeding, however, it is important to comment briefly about what formal strategic planning is *not*: strategic planning does not attempt to make future decisions. Decisions can be made only in the present. Forward planning requires that choices be made among possible scenarios and future events, but decisions made in their light can only be made in the present. Naturally, once made, these decisions may have long-term, irrevocable consequences.*

* According to Harold Koontz and Cyril O'Donnell: "Planning is deciding in advance what to do, how to do it, when to do it, and who is to do it. Planning bridges the gap from where we are to where we want to go. It makes it possible for things to occur which would not otherwise happen. Although the exact future can seldom be predicted and factors beyond control may interfere with the best-laid plans, without planning events are left to chance. Planning is an intellectual process, the conscious determination of courses of action. the basing of decisions on purpose, facts, and considered estimates." See Harold Koontz and Cyril O'Donnell, *Principles of Management: An Analysis of Managerial Functions*, 5th ed. (New York: McGraw-Hill, 1972), p. 113.

Do you long for the "good old days gone by" as you stare into the steely eyes of the future? Managers can no longer rest easy in the knowledge that today looks like yesterday and that tomorrow will be more of the same. Planning for the future requires managers to become trend spotters in an environment in which the globalization of markets and the pace and scope of change make it hard for companies to follow rapid shifts in technology, processes, and consumer taste. They must learn to identify discontinuous change and unsettling shifts arriving on an uncertain schedule. This provides an entirely new landscape for managers to navigate, and only the trend spotters can hope to conquer it.

The Achilles' heel of many organizations is often the inability to spot trends and act on them quickly. Michael Eisner, former CEO of Disney, said that if Disney "does not discern consumer trends, Tomorrow land could become Yesterday land before they know it."* Warren Buffett, the legendary investor, has a knack for describing such things articulately: "The rearview mirror is always clearer than the windshield."† Benjamin Franklin said it even more succinctly: "Look before, or you'll find yourself behind."

These insights show that identifying trends is a neat way to synthesize consumer behavior, help eliminate uncertainty, and identify new opportunities. In the 1980s, the most significant trends were increasing competition from all corners of the globe, greater demands for quality, shorter product life cycles, process improvement, and the need to create an agile work force. Most of these trends are now considered boilerplate, and experienced managers understand them quite well.

TREND OUTCOMES

Outcomes: Outcomes are the measured results that the organization realized as a result of the action taken.

So, what are the outcomes of the new trends today? What is going to be different about the next 10 years? Most obvious is the Internet. With 150 million people connected in only 15 years, the Internet has become the fastest

* Brainyquote.com
† Ibid.

accepted communications medium ever.* It took the telephone 70 years, radio 40 years, and television 15 years to reach that milestone. The Internet is no longer just a data network; it's a sales and distribution channel, and it's facilitated the ability to do business over the web. New e-business and organization master planning is enabling the integration of information industries—content, storage, networks, business applications, and consumer devices—that have been isolated until recently. This integration is leading to industry convergence: the melding of consumer electronics, television, publishing, telecommunications, and computers for the purpose of facilitating new forms of value. Clearly, the Internet groundswell is starting to affect every facet of our lives—personal and business.

TREND IMPLEMENTATION CHALLENGES

Trend spotting isn't just for entrepreneurs looking to start new companies or for marketers attempting to sell old products in new packages. It is useful for identifying new business opportunities as well. Consider, for example, the growing emphasis on well-being. This trend encompasses several mini-trends: the yearning for stress relief, a desire for greater balance in one's life, a revitalized interest in family and home, and a new focus on the environment. The business response: Grocery stores now stock natural and organic foods, medicinal herbs, and ready-to-eat meals. Insurance companies are beginning to cover alternative medicine. Travel agencies increasingly sell spiritual vacation packages. Hardware stores carry air and water purifiers and nontoxic paints.

Mastering the patterns of e-business requires understanding that the future will be built on a new set of operating assumptions. But first we must become expert in differentiating fads from trends. Fads catch on quickly, spread, and then die a fast death. Eileen C. Shapiro, in her iconoclastic book *Fad Surfing in the Boardroom*, defines fad surfing as "the practice of riding the crest of the latest management panacea and then paddling out again just in time to ride the next one; always absorbing for managers and lucrative for consultants; frequently disastrous for organizations."†

* A recent report by the US Department of Commerce estimates that the overall *data traffic on the internet* had been doubling every 100 days until the year 2000, and is currently doubling every 200 days or so.

† Eileen C. Shapiro, *Fad Surfing in the Boardroom* (Reading, Mass: Addison-Wesley, 1995).

In contrast, trends often start slowly but spread like wildfire as mass-market consumers and companies fan the flames with their pent-up demands. Trends are global, tend to last approximately 5 to 10 years, and may evolve dramatically. The web is a great example of a trend. It started slowly in 1989 in a remote lab in Switzerland, and with the advent of the Mosaic browser, burst onto the mass market, taking everyone by surprise. Did anyone you know identify the web as a major trend in 1994? The smart manager stands at the forefront of trends, such as the web, before they become mainstream topics. Since it takes years to steer large organizations in new directions, toward new horizons, an organization's captains must be aware of what lies ahead or their companies will sink as quickly as the *Titanic* when it hit the iceberg. Trend spotting has fast become a "plan or be planned for" issue.

In this chapter, we describe the major trends that are driving organizations to become e-business enterprises (see Figure 5.1). As a manager,

Trend Type	Trend Drivers
Consumer trends	1. Speed of service
	2. Self-service
	3. Integrated solutions, not piecemeal products
Service/process trends	4. Convergence of sales and service: customization and integration
	5. Ease of use: make service consistent and reliable
	6. Flexible fulfillment and convenient service delivery; streamline your supply chain
Organizational trends	7. Contract manufacturing: becoming brand intensive, not capital intensive
	8. Retain the core, outsource the rest: business process outsourcing
	9. Increasing process transparency and visibility: continuous innovation and employee retention
Enterprise technology trends	10. Enterprise applications: connect the corporation
	11. Infrastructure convergence: increasing melding of voice, data, and video
	12. Multichannel integration: computer telephony integration and voice recognition
	13. Wireless applications enter the mainstream
	14. Leveraging legacy investments: the rise of middleware for application integration

FIGURE 5.1
Major trends.

your ability to comprehend core trends will improve your chances of better grasping the opportunities facing your organization. As you read through this chapter, ask yourself, "What is the common thread running through these trends?"

Major Trends Affecting the Organizational Master Planning*

As time quotas shrink, customers look for companies that serve them faster. Just take a look at new trends in online car purchasing and home refinancing. The message to the marketplace is clear: To succeed, companies must reduce the processing time between search, selection, order entry, and order fulfillment. Delays at any step of the process are unacceptable! But why do delays occur? Often they're due to excessive handoffs.

Consider the case of a specialty stainless steel producer that wanted to improve its unacceptable 40% on-time delivery record. The organization identified unnecessary handoffs as delaying the production process. For example, each order was entered into the system three times. First, customer service would enter the information after writing down the buyer's specifications and use printed lists to check whether the order could be produced. They then checked printed schedules to determine a ship date. Second, operations would verify whether a particular grade of steel could be produced; information was then entered into the operations system. Third, production control would use its own files to verify and then reenter the information.

This repetition caused significant delays and errors. To solve the handoff problem, companies are investing billions of dollars in integrated systems, which is exactly what the stainless steel producer did. After taking a close look at its problem, the organization decided one possible solution was an integrated system for most of its business operations: accepting orders, triggering receivables, sending orders to production, sending requisitions to the warehouse, updating inventory, updating accounting, and replenishing stock with suppliers.

* "Trend Spotting: Anyone Can Play." *Businessweek*, March 2, 2008; updated March 2008. The article points out that trends often start slowly but spread like wildfire as mass-market consumers and companies fan the flames with their pent-up demands. Trends are global, tend to last approximately five to ten years, and may evolve dramatically. The Web is a great example of a trend identified in this article. It started slowly in 1989 in a remote lab in Switzerland and, with the advent of the Mosaic browser, burst onto the mass market, taking almost everyone by surprise.

What does this trend mean for e-business? When one considers the challenge of meeting the demands of busy, time-starved, dissatisfied consumers in an environment of hostile competition, low margins, and countless sales outlets selling similar products, it becomes clear that changing the entire business model is the only plausible strategy. Through an effective Organizational Master Plan and organizational alignment, the organization must cut the time customers wait for service. Customers now penalize companies that infringe on their time through delays, mistakes, or inconveniences. If companies don't expedite processes, customers will go to someone who does it faster. If one organization doesn't make it easy for the customer to do business, another will.

It is very important that managers understand and diagnose the cause behind service delays. They need to analyze if an integrated system can speed service. If so, they need to strategize, design, and implement such systems as soon as possible. Unfortunately for some companies, their managers may wake up too late to heed the sound of customers' voices screaming for faster service.

Unquestionably, the opportunities for digital revolution are many and largely unexplored. How should an entrepreneurial manager begin? By looking for ways to anticipate consumer trends, foresee technology trends, and envision new organizational forms. A lot of what we find surprising and unpredictable in today's environment is actually a series of events that has been played out in pretty much the same way in industry after industry. Once we see the pattern, we can understand and predict change—and begin to master it.

The starting point for all strategic planning cycles is a list of assumptions that the plan will be based upon. The excellence of these assumptions is largely based upon how well your organization can identify and analyze trends.

> Using commercially available technologies will not give you an advantage, but not using them will surely give you a significant disadvantage.
>
> **—H. James Harrington**

6

The Business Planning Process[*]

> Angel investors believe that the organization's leaders are more important than a new concept.
>
> —**H. James Harrington**

OVERVIEW

Though Business Plans have many different presentation formats, they typically cover five major content areas. At the end of this chapter, we offer a sample Business Plan outline format. The following are the main components of a well-thought-out Business Plan, and we also offer a summary of Business Plan pitfalls at the end of this chapter:

- Background information for appraising your current position
- Top management and board of directors have a 'strategic vision' at the core
- A marketing plan that helps you think competitively
- A summary of the operational plan, viewed as a production system
- A well-thought-out financial plan

[*] H. James Harrington and Frank Voehl, *The Organizational Alignment Handbook: A Catalyst for Performance Acceleration* (in press).

COMPONENT 1: APPRAISING YOUR CURRENT POSITION*

It is often told that the hardest part of initiating or appraising the current situation leading to the development of the Business Plan is determining an accurate assessment of your current position today. The reason is that it is not always as easy and obvious as we may think, as in the case of the Domino's Pizza chain. The key question to begin with is: What business are we in? The obvious answer is the "we sell pizza," but so do hundreds of competitors.

To try and differentiate the organization, the Domino's master planners decided that to position its products based on higher quality was too difficult to achieve, given its scarce resources available at the time. Accordingly, they recognized that the business Domino's was really in was the convenience industry, not the pizza business. Based upon its feedback scores, the pizza was not any better or any worse than most of the competition.† However, the niche Domino's chose for itself in its plan was the business of selling convenience: "Your pizza in 30 minutes or its free!"‡ Wow, what a breakthrough concept at the time, circa 1980. That's just the kind of penetrating thought that goes into assessing your organization's current position.

This particular section of the Business Plan is most appropriate for those who are developing a plan that is being used to seek financing. Within this section you will start by describing what stage of development your company is in and what the financing, if obtained, will be used for. Keep in mind that there are three basic reasons for seeking outside financing: (a) start-up financing, (b) expansion financing, and (c) work-out financing.

---

* Business appraisal and valuation is a relatively new field that has evolved quickly, driven especially by the corporate accounting scandals. As with any young discipline, there are growing pains. There are a wide range of people performing business valuations; some are accredited, some aren't. Even among the groups that accredit these, there is a huge range of experience required for the accreditation. In the practice of law, accounting, and medicine, there are certifications required by law to enable one to practice. That is not true with respect to business valuation. There is no business valuation–specific college major, and there are no requirements that must be met for one to claim to be a business appraiser. As a result, there are many who promote and perform business valuations without the knowledge and experience requisite of an expert, in many instances without being even minimally qualified.

† "Finding a Path through the Research Maze," by Patricia M. Cole. *The Qualitative Report* 2(1): 1994.

‡ Georgia Trial Lawyers Association, *Jury award spurs Domino's to drop deadly policy*, (2002), http://www.gtla.org/public/news/dominos.html (accessed September 18, 2007).

- **Start-up financing:** If your organization is in the process of seeking start-up financing, you will need to list specific milestones that have already been achieved and then you must emphasize all positive developments without stretching the truth and being misleading. The following are some of the key questions that lenders or investors may ask.
  - Has the market research been thoroughly done? What does it show? How do you plan on using it to move forward?
  - Has a prototype product been developed or is one in the works? When will it be ready?
  - Have facilities already been leased or purchased? For how long and for how much money each month?
  - Is the management team in place? Have they invested their own money in the organization? Who are they and what is their pedigree? Were they successful at operating other organizations? What types of contracts do they have? How about "golden parachutes"?
  - Has manufacturing been contracted for?
  - Have the initial marketing plans been finalized? Who is the competition? How can they be neutralized?
  - What is the makeup of the Advisory Board? Have they been successful with other organizations that are at the same point in their development?

  Whether you receive financing and the terms of that financing will depend upon the answers to the above questions and the stage of development that your organization is in. The more fully developed it is, the more favorable your financial arrangements will be.
- **Expansion financing:** If your business is already established as up and running and you are seeking expansion financing, you need to give clear evidence that you are not merely seeking financing as a way to solve existing cash-flow problems, or to cover losses for extraordinary expenses that many times are experienced during the initial start-up.
- **Work-out financing:** Many angel investors and banks/lenders do not like to offer work-out financing, and those who are willing to do so will want to see a business plan that clearly identifies the reasons for current or previous problems and provides a strong plan for corrective action. No matter what manner or type of financing you are seeking, lenders expect to be apprised of the source and amount of any capital that has already been secured, spent, or dissipated.

They also often expect key executives to have made substantial personal equity investments in the business. They will feel even more comfortable if they recognize any other investors who may have participated in earlier stages of the financing process.

COMPONENT 2: A STRATEGIC VISION IS THE CORE OF YOUR BUSINESS PLAN*

In almost every case, the first 50% of the Business Plan is geared toward helping develop and support a solid business strategy. Management and the board of directors must begin by looking at the market, the industry, the customer base, and its competition. They need to analyze and study customer needs and the benefits of current products and services. Next, they need to evaluate the strengths and weaknesses of each competing organization in order to look for opportunities in the marketplace during the next year and beyond. All of these activities are largely aimed at helping senior management create an ongoing strategy for the organization and its subsidiaries, if they have any.

The second 50% of the Business Plan is devoted largely to executing the selected business strategy. The products and services, marketing, and operations should all closely tie in with the strategy. So while it may be easy to select a smart-sounding strategy for your plan, we recommend you give a lot of thought to the *strategic vision* that will set the course for your business. Although some form of strategic planning is critical to achieving results, we maintain that the "old" strategic planning maps and templates are no longer viable because we are largely sailing through uncharted waters in a turbulent world, making it virtually

* Definition of *business strategy*, from Entrepedia: The entrepreneurship wiki. Formulating a strategy is really about deciding what the objectives of the business are, and developing an overall strategy for how they should be achieved within the business environment. This should be done in a way that best takes advantage of the company's resources and core competencies (what it does best—its key competitive advantages). Strategy formulation then needs to follow a process that includes: (1) understanding where the business is now: externally—in terms of its markets, customers, competitors and general business environment—and internally, in terms of its resources and competencies, i.e., swot analysis), (2) outlining objectives for the business in terms of where it wants to be—in order to continue towards (and hopefully also to thrive and grow in) the future, i.e., vision and mission statements, and (3) developing strategies and tactics for the business as a whole, and the different functions within it, in order to achieve these objectives, i.e., master planning strategy. www.entrepedia.org (accessed November 10, 2011).

impossible to predict long-term changes in the future for a system. Answers and direction emerge and a manager must be aware of them. In doing so, two major issues must be analyzed, evaluated, and dealt with: (1) the role of strategic vision in an unpredictable world and (2) the need to contrast complexity-oriented perspectives with traditional management perspectives.

Defining Strategic Vision: The Core of the Business Plan

Strategic vision is becoming a widely used term, but not well understood as many leaders don't strategically understand what vision is, or why it is important. One strategic leader is quoted as saying, "I've come to believe that we need a vision to guide us, but I can't seem to get my hands on what 'vision' is. I've heard lots of terms like mission, purpose, values, and strategic intent, but no-one has given me a satisfactory way of looking at vision that will help me sort out this morass of words. It's really frustrating!" (Collins and Porras 1991).*

One definition of a strategic vision comes from Burt Nanus, a well-known expert on the subject. In his writings, Nanus defines a strategic vision as a realistic, credible, attractive future for [an] organization.† Let's break this down as follows:

- **Realistic:** A strategic vision must always be based in reality to be meaningful for an organization, as opposed to an ordinary vision. For example, if you are a master planner charged with developing a strategic vision for a computer software company that has carved out a small niche in the marketplace developing instructional software and has a 3% share of the computer software market, a vision to overtake Microsoft and dominate the software market is not realistic, but a strategic vision might call for a 5% niche during the next 24 months.

* James Collins and Jerry Porras, "Organizational Vision and Vision Statements," *California Management Review*, 34(1), pgs. 30–52.

† Burt Nanus is a well-known expert on leadership and the author of many books on the subject, including *Visionary Leadership.* Now professor emeritus of management at the University of Southern California, he was also research director of the Leadership Institute. His most often quoted comment on the subject of strategic visioning is: "[Strategic vision] is where tomorrow begins, for it expresses what you and others who share the vision will be working hard to create. Since most people don't take the time to think systematically about the future, those who do, and who base their strategies and actions on their visions, have inordinate power to shape the future." *Leaders Who Make a Difference: Essential Strategies for Meeting the Nonprofit Challenge*, by Burt Nanus and Stephen M. Dobbs (San Francisco, CA: Jossey-Bass, 1999), 279–426.

- **Credible:** A strategic vision must be believable to be relevant. More importantly, it must be credible to the employees and all members of the organization. If the members of the organization do not find the vision strategically credible, it will not be able to create and harvest energy, or serve a useful and meaningful purpose. One of the inherent purposes of a strategic vision is to inspire those in the organization to achieve a level of excellence, to redouble their energy, and to provide purpose and strategic direction for the employees; to do so it must be deployable.* A vision that is not credible or deployable is not strategic in its nature and will accomplish neither of these ends.
- **Attractive:** If a vision is going to strategically inspire and motivate those in the organization, it must have a strong magnetic attraction and draw people to it in order to create energy and synergistically leaven the organizational energy. In other words, people must want to be part of the future that's envisioned for the organization.
- **Future based:** A strategic vision is never in the present, and is always in the future, yet is somehow tied to the past. In this respect, the image of the leader and his planners gazing off into the distance to formulate a vision may not be too far from the truth, but can be somewhat misguided if not channeled by the core values of the organization. The paradox lies in the strategic vision, which is not where you are now, but where you want to be strategically in the future, providing you do not repeat the failures of the past; for otherwise, you are organizationally doomed to repeat it.

Nanus in his writings goes on to say that the right strategic vision for an organization—one that is a realistic, credible, and an attractive future for that organization—accomplishes a number of things for the organization:

---

* Policy Deployment (aka) *Hoshin Kanri* is a method devised to capture and cement strategic goals as well as flashes of insight about the future, and develop the means to bring these into reality. Also called *policy deployment* or *hoshin planning*, it is a business planning methodology that uses a plan, do, check, act (PDCA) cycle to create goals, choose control points (measurable milestones), and link daily control activities to company strategy, and is covered in detail in *The Organizational Alignment Handbook: A Catalyst for Performance Acceleration*, by H. James Harrington and Frank Voehl (in press). "Hoshin Kanri (also called Policy Deployment) is a method for ensuring that the strategic goals of a company drive progress and action at every level within that company. This eliminates the waste that comes from inconsistent direction and poor communication. Hoshin Kanri strives to get every employee pulling in the same direction at the same time. It achieves this by aligning the goals of the company (Strategy) with the plans of middle management (Tactics) and the work performed by all employees (Operations)." http://leanproduction.com/hoshin-kanri.html. Akao, Yoji, ed. (Japan, 1988; English, 1991) Hoshin Kanri, *Policy Deployment* for *Successful* TQM (New York: Productivity Press; originally Japanese Standards Association), xiv.

- It attracts commitment and energizes people. This is one of the primary reasons for having a vision for an organization—its motivational effect. When people can see that the organization is committed to a vision and that it entails more than just having a vision statement, it generates enthusiasm about the course the organization intends to follow, and increases the commitment of people to work toward achieving that vision.
- It creates meaning in workers' lives. A vision allows people to feel like they are part of a greater whole, and hence provides meaning for their work. The right vision will mean something to everyone in the organization if they can see how what they do contributes to that vision. Consider the difference between the hotel service worker who can only say, "I make beds and clean bathrooms," to the one who can also say, "I'm part of a team committed to becoming the worldwide leader in providing quality service to our hotel guests." The work is the same, but the context and meaning of the work are different.
- It establishes a standard of excellence. A vision serves a very important function in establishing a standard of excellence. In fact, a good vision is all about excellence. Tom Peters, the author of *In Search of Excellence* (1988), talks about going into an organization where a number of problems existed. When he attempted to get the organization's leadership to address the problems, he got the defensive response, "But we're no worse than anyone else!" Peters cites this sarcastically as a great vision for an organization: "Acme Widgets: We're No Worse than Anyone Else!" A vision so characterized by lack of a striving for excellence would not motivate or excite anyone about that organization. The standard of excellence also can serve as a continuing goal and stimulate quality improvement programs, as well as providing a measure of the worth of the organization.
- It bridges the present and the future. The right vision takes the organization out of the present and focuses it on the future. It's easy to get caught up in the crisis of the day, and to lose sight of where you were heading. A good vision can orient you on the future and provide positive direction. The vision alone isn't enough to move you from the present to the future, however. That's where a strategic plan, discussed later in this chapter, comes in. A vision is the desired future state for the organization; the strategic plan is how to get from where you are now to where you want to be in the future.

Another definition of vision comes from Oren Harari*: "Vision should describe a set of ideals and priorities, a picture of the future, a sense of what makes the company special and unique, a core set of principles that the company stands for, and a broad set of compelling criteria that will help define organizational success."

An additional framework for examining vision is put forward by Collins and Porras. They conceptualize vision as having two major components: a guiding philosophy, and a tangible image. They define the guiding philosophy as "a system of fundamental motivating assumptions, principles, values and tenets."† The guiding philosophy stems from the organization's core beliefs and values and its purpose.

A strategic vision expresses core values that inform action, and it nurtures the capabilities needed to pursue this action. It is so very much more than speculating about future trends and events or forecasting the impact of future technologies. It is grounded in the past while looking to the future. And because such a strategic vision is durable and long-lasting, it should not require revision every few years like most ordinary visions do. The following example shows that while the concept may be hard to explain, you will almost always know it when you see it.

> Take the case of America's founding; the "strategic vision" that helped form and create the United States of America's original thirteen colonies is embodied in the Declaration of Independence (vision), the Constitution (strategy), and the Bill of Rights (values). With these three documents, the nation's founding fathers drew on an Old World heritage tempered by New World experiences. Specifically, the Declaration's concepts of natural rights were rooted in the Enlightenment period, and its revolutionary political ideals were forged by life in Colonial American period. This one-page document, which is timeless and is still relevant today, expresses values that placed man in a unique and revolutionary position with respect to the government.
>
> The Constitution was key because it initially translated the Declaration's ideals into a practicable reality by developing appropriate political

* Oren Harari is a professor at the Graduate School of Business in the University of San Francisco where he teaches strategic and global management to really sharp, talented MBA and executive MBA students; several of them work with him in doing cutting-edge research on the most successful organizations and leaders of today. He was honored to be chosen and interviewed as one of the 40 "best minds" in management in the world by the editors of *Business Minds*, published by the *Financial Times (London)* in fall 2001. Over the past 25 years, he has spoken and consulted with high-caliber audiences and clients all over the world, and some of these groups are listed in the "Speaking" section of his website, which is found at http://www.harari.com.

† James Collins and Jerry Porras, "Organizational Vision and Vision Statements," *California Management Review*, 34(1), pgs. 30–52.

instruments that were effective for a democratic republic. The fact that it has had, in the course of over 200 years, only 27 amendments attests to its remarkable durability and strategic workability. Together, the Declaration's expression of core values and the Constitution's pragmatic translation of those values into capabilities helped to set a young nation America on a radical new course with a powerful strategic vision to guide it.

The Role of the Quality Policy and Quality System

We are seeing that the quality policy and its attendant quality system are vital parts of the Business Plan for large organizations and for many small businesses as well. Also, large organizations are demanding that their suppliers can demonstrate consistent quality in the delivery of products and services. As a result, small organizations are realizing that one way to win and retain more business is to provide customer satisfaction through quality service, which is articulated in the *quality policy* and operationalized in the *quality system.*

There is no universal agreement on a definition of *quality.* In many organizations it is understood in terms of meeting customer requirements or lack of defects. It is also sometimes described as "delighting the customer by fully meeting their needs and expectations." However, there are recognized quality standards for performance—such as the Baldrige*, International Organization for Standardization (ISO) 9000, TQM,† and the Japanese Management System (JMS)—as well as a range of quality management systems, benchmarking schemes, and Business Excellence

* The Malcolm Baldrige National Quality Award was the first program of this kind in the United States designed to help improve quality and productivity. This is accomplished by helping to stimulate American companies to improve quality and productivity for the pride of recognition while obtaining a competitive edge through increased profits. Its focus is recognizing the achievements of those companies that improve the quality of their goods and services, and by providing an example and a "living lab" to others. It does this by establishing guidelines and criteria that can be used by business, industrial, governmental, and other organizations in evaluating their own quality improvement efforts. Providing specific guidance for other American organizations that wish to learn how to manage for high quality is accomplished by making available detailed information on how winning organizations were able to change their cultures and achieve eminence. Baldrige winners are also required to hold visitor orientations, which are one-day open learning events at the organizational headquarters, usually offered free of charge.

† Total Quality Management (TQM) is an approach to improving the competitiveness, effectiveness, efficiency, and flexibility of an organization for the benefit of all stakeholders. The implementation of TQM depends on two key factors—commitment and leadership. Given these two values, successful implementation of TQM should result in a company with satisfied customers, co-operative suppliers, and profitable performance. Unfortunately, TQM was often viewed as a failure, primarily due to low levels of rapid results and the failure rate of implementations has often been measured at 90% or greater.

Models (BEMs)*, some of which are outlined in this book and in our other writings on the subject. For the small business, the thought of introducing a formal quality system may seem daunting. It is also the case that potential customers will not always require you to have achieved a particular quality standard. But most of your customers will expect you to have some quality measures and a basic quality system in place—and it doesn't have to be that complicated if you take it on a step-by-step basis, using our Organizational Master Plan as a guide. And the best part of all is that having a quality system in place makes it easier for you to run your organization more efficiently, enhance your reputation, and avoid costly mistakes.

What Do I Need to Measure?

Your measurement system should basically express the standard that you set for the key functions and outputs within your organization, and then show how you will provide measurement that the standard is achieved. Examples of the standard and the measure are

- **The standard:** We will send all potential customers a written estimate for their job request and specifications within 48 hours of receiving their enquiry.
- **The measure:** All enquiries will be logged into our computer system as they are received and all estimates will be logged for the time they are dispatched. Every five days, the planner or estimator will run a data query that compares date received with date dispatched and a summary report will be submitted to the section supervisor.

In this example, your measurement system should be able to tell you and your customers how well your business is performing and to reveal where it is not reaching the required standard so that you can take remedial action. Performance measures can be applied to every area of your business operation, but to get started it makes sense to concentrate on

* The Business Excellence Model (BEM) offers a means for companies to measure their performance on the world-class scale and to focus on the needs of their customers. It can be used as a businesswide framework in a holistic, focused, and practical way, and can offer significant value to Business Plans in global organizations. A cornerstone of the BEM approach is to recognize that the processes at work in an organization are the means of harnessing and releasing the talents of its people.

those areas that directly impact on the goods or services that you provide to your customers.

Designing Your Organization's Quality System

Attaining a recognized quality accreditation, such as Baldrige, Sterling,* ISO 9000,† or Investors in People‡ may well be your long-term aim, so you want to be clear that the quality systems that you introduce are going to help you achieve that ambition. Unless you are experienced in quality management, you may want to seek some professional advice. In any case, it is important that your quality measures are "sensible" and appropriate to your business. They should add to your efficiency and success and not prove a drain on your time and resources.

The quality policy should encourage your master planners and managers to think about the important issues:

- Do we deliver on time?
- Do we deliver on budget?
- Are our customers pleased with our service?
- Did it perform to the customer's requirements?

* The Sterling is the Florida Sterling Award, which is one of the more well-known and popular of the state quality awards in the United States. Of the 52 states, for example, more than half have their own version of the Baldrige performance management system. Since 1992, the Florida Sterling Council has assisted organizations in utilizing proven standards of excellence as a guide to making improvements that generate better operations, customer value, and overall results. The Sterling Award has the following objectives for client organizations: (1) Improve overall performance and the bottom line, (2) accelerate improvement efforts, (3) exceed customer and stakeholder expectations, (4) identify, focus, and engage all employees on a common set of goals, and (5) capitalize on today's ever-changing environment.

† ISO 9000 is an offering of the International Organization for Standardization (ISO), which is a worldwide federation of national standards bodies from some 140 countries, one from each country, established in 1947 as a nongovernmental entity. The ISO 9000 family is a set of international quality standards and guidelines that has earned a global reputation as the basis for establishing quality management systems. The main advantage of this family of standards is that it enables a company to compare its system with others. Any kind of organization can apply for ISO 9000 registration, whether a manufacturer, service industry, retailer, charity, educational, professional or public sector body. To achieve ISO accreditation, you will probably need professional support from a quality consultant such as the Harrington Institute or Strategy Associates.

‡ Investors in People (IiP) is the national standard that sets a level of good practice for training and development of people to achieve business goals. IiP is designed to help businesses achieve continuous improvement in performance and competitiveness. It does this through a planned approach to setting and communicating business objectives and developing people to meet these objectives. With its strong people focus, IiP ensures that people are able and motivated to do what their organization needs them to do.

Then consider what evidence you will need to prove and improve your performance levels. For example, a simple customer questionnaire presented and collected at the end of every job or on delivery of your products can be very effective. With even these basic measures, you will have the beginnings of a quality system in place and a sound platform to build on.*

The Business Case for Quality

A well-designed and well-managed quality system should deliver ongoing and measurable benefits for your business, some of which should be spelled out in the Business Plan, especially if the organization has been in operation for three years or longer. These can include

- Improved customer satisfaction
- Stable and growth-oriented customer base
- Improved reputation leading to more repeat and follow-on business
- Early identification of problems leading to speedy resolution
- Reduced costs by not having to rectify mistakes through rework or recalls
- *Right first time* efficiency and *quality at the source*
- Improved efficiency and more rapid results through a working-smarter mindset
- Greater job satisfaction and worker motivation and self-control

Managing Your Organization's Supply Chain

- **Supplier Controls:** These are the preventive measures that are put into place to minimize the possibility of suppliers providing unacceptable product. They include things like supplier qualification, requirements placed on the supplier to be ISO 9000 certified, source inspection and receiving inspection,
- **Supply chain management:** Supply chain management is the creation of a management process for integrating decisions, plans, and information systems from customer requirements through the fulfillment process to the suppliers of materials.

---

* H. James Harrington and Frank Voehl, *The Organizational Alignment Handbook: A Catalyst for Performance Acceleration*, (in press).

To provide value-added products and service, the Business Plan for larger organizations should spell out how you will be managing the total supply chain. So much business is actually conducted verbally with no traceable evidence of agreements about delivery schedules, quality measures, the products to be used, and so on. By establishing clear contracts and progress-defining measures with your organization's internal processes and the external suppliers, you will be able to avoid confusions, misunderstandings, and mistakes. You will also be able to work more intelligently with your total supply chain helping everyone to understand your needs and creating lasting and reliable partnerships. Setting up a system to manage your supply chain may take time, and at first may seem to be a paperwork burden. However, you should soon see practical benefits through you and your suppliers having an improved knowledge of each other's requirements.*

COMPONENT 3: THINK COMPETITIVELY THROUGHOUT YOUR PLAN

In today's highly competitive and overly crowded marketplace, you're probably going to have serious competition no matter how creative your business concept. That is why you need to think competitively throughout your Business Plan. You need to realistically identify where you will do things in a manner similar to your competitors, where you will do things differently, where you have real strengths, and where you have real weaknesses.†

As part of the business planning process, you need to learn how to develop and write an effective marketing plan that answers the following questions:

1. How large is the potential market? How do you know? How is it segmented?
2. Does your customer analysis show that the is market growing, flattening, or shrinking? What data supports your conclusions?

---

* Harrington and Voehl, *Little Big Book*.

† While data is being collected, the analysis process can begin. In our research, we noticed how patterns formed from one interview to another. After the interviews were completed and transcribed, we began a formal coding system by organizing these patterns into central tendencies and ranges: (1) central tendencies describe how the data chunk together into the research participants' common themes or categories, (2) ranges allow for the differences within those categories to be discussed.

3. How many people or businesses are currently using one of your competitor's products that are the same or similar to the one you are offering or plan to offer? How can you differentiate yours? How many customers or prospects potentially have any possible use for the product?

Market Segmentation

In almost every case that we have studied over the past thirty years, every market has some major and distinctive segments. Even if the marketplace is not currently segmented, the probability that it could or will be is real and of urgent concern. This is particularly true if the marketplace for your product or service is multiregional or international. If this is the case, segmentation is vital, especially for a small organization, if your desire is to be competitive because your master planners will need to discuss segmentation within your business category. How you intend to cope with any positive or negative effects that segmentation may have on your particular business is vital, as most marketplaces are segmented by price, features, and quality issues. Generally, however, price and quality alone do not provide the most clear or definitive market segmentation. Much stronger segmentation can usually be found through an evaluation of product features or service uses, and their importance to various consumers.

Consumer Analysis

In your Business Plan, you will need to evaluate the typical consumers within the market segments you are targeting in order to determine if the market is growing, flattening, or shrinking. Although there are innumerable variables to consider when analyzing consumer behavior, you need to try and develop surveys to focus on those behavioral possibilities that best determine how viable your product will be in your target markets.*

The customer analysis starts by looking at which features will most appeal to consumers.

* What is a survey? The American Statistical Association (ASA) discusses current practice and the innovations now in process for the main types of surveys—mail, telephone interview, and in-person interview. Each form has distinct advantages and disadvantages. Each is also being greatly changed by the use of computers, the Internet, and telecommunications.

- How are choices made between competing products and which marketing promotions or media avenues seem to offer the best vehicles for reaching the consumer base?
- How much disposable income do target consumers have to spend on this product?
- How do your target consumers reach purchasing decisions?
- Are consumers presold on a particular brand before they visit a store or do they buy on impulse?
- What characteristics influence the purchase of one product or service over a competing one?
- How many people or businesses are currently using one of your competitor's products that are the same or similar to the one you are offering or plan to offer.

You will need to include an overview of the organizations and their products, programs, and services with which you will be in direct competition. Next, try to identify the market leader and define what makes it successful and emphasize those characteristics of the firm or offerings that are different than yours. The initial reaction is to dismiss this section of the Business Plan simply because you may not have any real current competition. However, if there isn't a product or service similar to yours on the market, then you need to identify those firms that provide products or services that perform essentially the same or similar functions by making an attempt to identify any organizations that are likely to enter the market or those competitors that are in the process of developing products or services that will be competitive with those your organization is or will be offering. It is very important to be clear not only about the distinguishing features of your program, product, or service, but also to delineate any strong consumer benefits. What makes your product or service significantly better than competitive offerings?

Next, you will need to do an in-depth analysis of the competitive advantages and weaknesses of your organization and should include information that will help allay any concerns that may arise as to their ability to significantly hinder your success. This particular section of the Business Plan is important, especially if your company is a start-up, because you will be competing with established companies that have inherent advantages such as brand recognition, financial strength, and established distribution channels.*

Additionally, strategic positioning can be thought of as a marketing strategy for your product or service because it defines how you are going to feature your product to your targeted marketplace. The first step here is to decide who your target market will be, consisting of those potential customers toward whom you will direct most of your marketing efforts. In many cases, this group will not be the unique or even the largest market for your product, but based on competitive factors and product benefits, it is the one that you feel your marketing arm can most effectively reach. Start-ups are generally successful if they focus on a highly specific, yet very narrow target marketplace, as big general markets are usually dominated by large, well-established organizations.

Once you have determined who your target market is, you need to decide if you want consumers to perceive your product as the premium quality leader or as a low-cost substitute. Also, if you have a one-product organization or more likely a service company, your marketing strategy may coincide with your overall business strategy. This doesn't necessarily have to be the case, but it is important that your product strategy be in sync with your overall business strategy.

Advertising and promotional activities are used to provide an overview of your general promotional plan, and you need to provide a summary of what methods and media you intend to use and why.† Although an

---

* A broad definition of *competitive analysis* is the action of gathering, analyzing, and distributing information about products, customers, competitors, and any aspect of the environment needed to support executives and managers in making strategic decisions for an organization. Key points of this definition include: (a) Competitive intelligence (CI) is an ethical and legal business practice. This is important; CI professionals emphasize that the discipline is not the same as *industrial espionage*, which is both unethical and usually illegal. (b) The focus is on the external business environment. (c) There is a process involved in gathering information, converting it into intelligence, and then utilizing this in business decision making. CI professionals emphasize that if the intelligence gathered is not usable (or actionable), then it is not intelligence. From Wikipedia; for further information, see http://en.wikipedia.org/wiki/Competitive_intelligence.

† On the website of the Department of Measurement, Statistics, and Methodological Studies (MSMS), College of Education, Arizona State University, you will find Dr. B's Wide World of Web Data with links to hundreds of online datasets all over the world that students and professors can use for in-class work. The Data Gallery provides a large number of graphic depictions of data. The Equation Gallery describes and presents various formulas for the statistics in one and in two dimensions. In the Graphic Studio you can experiment with data entry forms that produce plots in HTML. In the Computing Studio you can type data on a form that generates a page that walks you through the activities of computing your statistics. And just for the fun of it, you can get a random statistical quote from a collection of quotes from the history, philosophy, and practice of data analysis and statistics. It might stimulate your reflection on common statistical practice. Website editor: John Behrens.

advertising slogan or unique selling proposition isn't strictly necessary, you should outline the proposed mix of your advertising media, use of publicity, and/or other promotional programs as follows:

- Explain how your choice of marketing strategies and tools will allow you to reach your target market.
- Explain how they will enable you to best convey your product features and benefits.
- Be sure that your advertising, publicity, and promotional programs sound realistic, based upon your proposed marketing budget.
- Effective advertising generally relies on message repetition in order to motivate consumers to make a purchase.
- If you are on a limited budget, it is better to reach fewer, more likely prospects than too many people occasionally.

Your sales strategy needs to be in harmony with your business strategy, marketing strategy, and your company's SWOTs (strengths, weaknesses, opportunities, and threats). If your start-up company is planning on selling products to other businesses in a highly competitive marketplace, your marketplace entrance unveiling will be easier if you rely on wholesalers or commissioned sales representatives who already have created and formed an established presence and reputation in the marketplace. On the other hand, if your business is to sell high-tech products with a range of customized options, your sales force needs to be extremely knowledgeable and personable.

Your executives need to remember that banks generally lend money to businesses on a short-term basis, and many venture capitalists may want to cash out in a few years.*

* See R. J. Chenail, "Qualitative Research: Central Tendencies and Ranges." *AFTA Newsletter* (Spring 1992): 43–44. B. G. Glaser and A. L. Strauss (1967). *The Discovery of Grounded Theory: Strategies for Qualitative Research* (New York: Aldine De Gruyter, 1967). Y. S. Lincoln and E. G. Guba (1985). *Naturalistic Inquiry* (Beverly Hills, CA: Sage, 1985). A. L. Strauss and J. Corbin, *Basics of Qualitative Research* (Newbury Park, CA: Sage, 1990).

COMPONENT 4: THE OPERATING PLAN AS A PRODUCTION SYSTEM

Think of your Business Plan as a production system with the inputs (what goes in at the start) are raw material and unfinished assemblies.* In our case, the raw materials include

- Talent and initiative from your employees
- Capital
- Market position
- The company's creditworthiness
- The organization's earning capacity
- Assessment of changes in the marketplace

In the master planning process, the structure of the Business Plan is unique and in many cases, like nothing you've seen before. This part of the Master Plan is a nuts-and-bolts document, how-to-get-it-done-without-fail tool kit designed to grab your organization and take it where it needs to go. You won't find much of the theoretical conceptualizing and strategizing so many other planning books carry on about. That works for larger companies that can wait five years or more for results. Instead, what happens to your organization during the next twelve months concerns the Master Plan, for if you can hit your short-range targets year after year, the longer term will take care of itself.

In the Business Plan, operations is a catchall term used to describe any important aspects of the business not described elsewhere, and should be covered in a Annual Operations Plan, which is covered in detail in another

* According to Strauss & Corbin (1990), following data analysis, the challenge of the "operating plan's production system" is how to transform these materials clearly and effectively so that others can benefit from them. The *Chenail Qualitative Matrix*, developed at Nova Southeastern University (NSU), insures a relationship between data presentation, data analysis, and the literature review. The matrix not only provides a conceptual frame for coding the data, but also suggests a map for reproducing analyzed data into an organized pattern that connects the findings of the research with the review of the literature. The terms *expected* and *unexpected* are used to organize the data presentation: (a) *Expected* refers to data that confirms the ideas of the authors in the literature review or the researcher's assumptions/ (b) *Unexpected* refers to data that departs from the authors' ideas in the literature review or the researcher's assumptions. Paricia M. Cole, "Finding a Path through the Research Maze," *The Qualitative Report* 2, no. 1 (1994).

section of this book. If the start-up is a manufacturing organization, you will need to discuss critical elements of the manufacturing process. For Business Plans covering retail businesses, you need to discuss store operations, while wholesalers discuss warehouse operations. In addition to discussing areas that are critical to operations, briefly summarize how major business functions will be carried out, and how certain functions may run more effectively than those of your competitors.

If there are any key vital positions that have not been filled, you will need to describe position responsibilities and the type of employment experience necessary for the position. If there is a board of directors, introduce each member and provide a concise summary profile of each person's background. Also, if they will have an active role in running your business organization, you should elaborate on that role. Finally, if consultants have been engaged for key responsibilities, include a brief description of their backgrounds and functions, how they will be engaged, and their exit strategy. Finally, fill as many of your key positions as possible before you seek funding. Many financiers reject plans if the management team is incomplete.

Our Sterling Restaurant case study illustrates the basic business planning structure and identifies the various techniques and methods used in the process, using many tools found at the *Microsociological Software Gallery*, such as the Chenail Qualitative Matrix and other important strategic vision planning tools.* (The Chenail Qualitative Matrix insures a relationship between data presentation, data analysis, and the literature review. The matrix not only provides a conceptual frame for coding the data, but also suggests a map for reproducing analyzed data in an organized pattern that connects the findings of the research with the review of the literature). The terms *expected* and *unexpected* are used to organize the data presentation: By the end, you'll have an idea of how each part of

* The purpose of the Microsociological Software Gallery site is to further research and education in the field of computer-assisted learning. Its strategic vision intended to stimulate the sociological imagination. For example: (a) in the Animations & Tutorials section you can test your Attitude Consistency, learn about the mechanism of Causal Attribution, test how familiar you are with C. Wright Mill's theory of the power elite, and much more. (b) In Games & Stimulations you can play Coalition Formation, the Mixed Motive Game, or a Zero Sum Game. The Utility section contains diverse instruments such as an introduction to authoring your HyperCard Stacks, and Sivanga Yoga 1.0, which might help you to promote a clear mind. Much of this software has been developed by Spitzer and is freeware or shareware. Visit http://www.soc.umn.edu/~spitzer/softwaregallery.html for details. Paricia M. Cole, "Finding a Path through the Research Maze," *The Qualitative Report* 2, no. 1 (1994). http://www.nova.edu/ssss/QR/BackIssues/QR2-1/cole.html.

the master planning process builds on what went before.* Further, you'll understand how the Business Plan ties all the various functions in your business together. Master planning challenges people to develop strategies to cope with the many problematic situations they are likely to encounter when traveling, studying, or working abroad; it can also be very useful for promoting sensitivity to cultural diversity among peoples of differing sexes, ages, and ethnic identifications and who belong to the same culture-at-large, as the Microsociological Software Gallery explains and details.

COMPONENT 5: A WELL-THOUGHT-OUT FINANCIAL PLAN

In this section you need to show projected, or *pro forma*, income statements, balance sheets, and cash flow. Existing businesses should also show historical financial statements. While how far into the future you need to project and the number of possible scenarios you can anticipate depends upon the complexity of the business, three to five years for financial projections and three scenarios are average. Scenarios should be based on the most likely course your business will take, a weak scenario with sales coming in well under expectations, and a good scenario with projected sales well over expectations.

Pro forma income statements should show sales, cost of operation, and profits on both a monthly and annual basis for each plan year. For all but the largest businesses, annual pro forma balance sheets are all that are necessary. Cash flow pro formas should be presented in both monthly and annual form. And, if your business is already established, past annual balance sheets and income statements should also be included. Include information that will assist potential lenders in understanding your projections.

* In applying the Chenail Qualitative Matrix, researchers have an opportunity to discover the serendipitous or unexpected instead of staying focused only on what was known through literature searches and previous observations. As Chenail concludes: "Qualitative researchers have a habit of focusing on what is familiar and central to the study at hand. What may be missed through this study of inquiry is an opportunity for investigators to know what might not be known to them prior to the study. Also, the margins of a project often provide some of the most interesting and informative patterns for investigators if they include a curiosity for the exception in their work and a hesitancy to explain quickly that which might turn out to be unexplainable." See R. J. Chenail (1992, Spring). "Qualitative Research: Central Tendencies and Ranges," *AFTA Newsletter* (Spring 1992): 43–44.

Lenders will give as much credence to the assumptions your projections are based on as they do the numbers themselves.

BUSINESS PLAN PITFALLS*

The following are the twelve of the most significant pitfalls that most executives and master planners face in building and maintaining their Business Plans. While they are not inclusive, they represent a consensus of belief on this subject area.† (The reader is encouraged to add his or her own pitfalls as they see fit).

- **Focusing Too Much on the Future.** It is commonly accepted that investment potentials and valuations for organizations are based on an organization's projected future performance. However, since the best indicator of future performance is past performance, or a company's past track record, Business Plans need to clearly show what milestones and accomplishments the organization has already achieved. Prior year successes in achieving goals tend to give investors the confidence that the management team will be able to deliver in the future.
- **Making Financial Projections Too Aggressive.** Since many investors go directly to the financial section of the Business Plan, it is critical that the assumptions and projections be realistic. Plans that show per-employee figures for penetration, operating margin, and revenue that are poorly thought out and reasoned, or are internally inconsistent or simply unrealistic greatly damage the credibility of the entire Business Plan. In contrast, clear, well-reasoned financial

* In our experience of assisting more than 1,000 start-ups, small, middle-market, and Fortune 500 companies with their business plans, as well as government and various nonprofit organizations, we have noted the common business plan pitfalls outlined in this book.

† Over the past twenty years, the Harrington Institute and its business partner Strategy Associates have been engaged in hundreds of projects with senior executives, the result of which has been an emerging body of knowledge in the area of Master Planning for Success.™ These clients have represented all types of businesses and marketplaces around the world. Our many client engagements have enabled us to create and align the body of-knowledge for many of the management disciplines, such as quality management (Six Sigma, policy deployment, TQM, ISO 9000), project management (aligned with the Project Management Body of Knowledge [PMBOK]), knowledge management, and change management, to name a related few.

assumptions and projections communicate operational maturity and credibility. By accessing and basing projections on the financial performance of public companies in their marketplace, ventures can demonstrate that their assumptions and projections are attainable.

- **Not Including Successful Companies in the Competitive Discussion.** Too many Business Plans try to show how unique their organization is and do not demonstrate their knowledge of the competition, which usually has a negative connotation. If few organizations are in a marketplace, the implication is that there may not be a large enough customer need to support the venture's products and/or services. When positioned properly, including successful and/or public companies in a competitive space can be a positive sign, as it implies that the market size is big. It also gives investors some assurance that if the management team executes well, the organization has substantial upside profit and liquidity potential.
- **Asking Investors to Sign a Nondisclosure.** Many investors will not want to sign a Nondisclosure Agreement because a business's strategy and/or concept are typically not confidential. Although it is possible that a key partnership is confidential, for the most part, the execution of the strategy and concept is what will make the organization successful and a desirable investment. The thinking is that if the concept or strategy must remain confidential, it often implies that there are no barriers to competitive entry. On the other hand, if a competitor or competitors can quickly copy the concept, then the business model is probably flawed and not sustainable. Also, proprietary technology is confidential and the business plan should not discuss the confidential aspects of the technology, but should discuss the features and benefits of the technology, and how these benefits fulfill a large customer need. As a serious investor will review the actual technology during the due diligence process, discussions regarding the signing of a Nondisclosure Agreement would be more appropriate at that point.
- **Indiscriminately Incorporating Investor Feedback into Your Business Plan.** It is a well-known fact that investors have different tastes as one investor may love a concept and the business plan while the next may dislike both. It is important to understand that business plans are working documents and are always undergoing iterations. Management teams must not be anxious to incorporate each

potential investor's comments. Instead, it is usually wise to have several investors and other business colleagues review the plan and provide feedback, and then incorporate common concerns, and probe other comments to determine if they are valid.

- **Not Clearly Articulating the Organization's Quality Policy.** The failure to clearly articulate the organization's quality policy toward its stakeholders and its stakeholder groups is a significant weakness in many Business Plans.* Part of this problem stems from the misguided view of the Quality Policy as being "an ISO thing" and part from management's failure to clearly focus on and articulate in writing what the critical success factors really are for their marketplace and organization in particular.
- **Tailoring Management Team Biographies to the Venture's Development Phase.** The management team section should always include biographical profiles of key team members and explain their responsibilities. These biographies should be tailored to the organization's development life cycle stage, since different skill sets are needed to launch, grow, and maintain a company. A start-up organization should emphasize its management team's success in launching and growing similar ventures. On the other hand, a more mature organization will want to outline and emphasize how management team members have successfully operated within the framework of larger enterprises.
- **Overemphasizing First Mover Advantage.** A Business Plan must include strategies that demonstrate the venture can and will build long-term protective strategies around its customer base. Simply claiming a first mover advantage is not always compelling in today's funding environment, and the methods through which the company will retain customer base should always be detailed in the business plan. Such methods could include implementing customer relationship management (CRM) tools, building networks (e.g., the more

* An organization should have an *quality policy*. This policy is meant to be an outward statement of its commitment to its customers and to the provision of quality products. The quality policy is a more abstract concept. It finds its practical and measurable expression in the quality objectives. The quality policy should be in line with the purpose and mission of the organization or company for which it is written. It also should be understandable by all members of the organization, employees, managers as well as stakeholders. One of the accepted quality policy definitions offered by ISO 9000:2000 is "The overall intentions and direction of an organization towards its stakeholders in relation to quality, as formally expressed by top management."

people that use the product or service, the harder it is for a competitor to penetrate the market), and describe ongoing value-added services.

- **Focusing Too Much on the Company's Proprietary Technology.** Since an organization's proprietary technology is a significant factor in investment decisions, it is much more important to show how this technology satisfies a large, waiting-to-be-filled customer need. Many companies fail because they do not understand the needs of their customers; understanding true customer wants and needs, while identifying which target markets most exemplify these needs, helps to outline a plan to penetrate these markets, which are critical to funding and execution success.
- **Defining the Market Size for a Company Too Broadly.** Defining the market size for a venture too broadly provides no value for the investor and is often viewed in a negative way. For example, mentioning the size of the US healthcare or global electronics markets is generally extraneous since no company could capture a meaningful percentage of either market. Rather, a more meaningful metric is the relevant market size, which equals the company's sales if it were to capture 100% of its specific niche of the market. Defining and communicating a credible relevant market size and a plan to capture a significant share within this market is far more powerful and believable to investors.

If this future is unpredictable, then an organizationwide "shared vision" is oftentimes a pipe dream and almost impossible to formulate. Such a vision in many cases is an illusion or interpretation made with hindsight, while vision and long-term plans become fantasy defenses against anxiety. Instead, managers need to learn to focus on ever-changing agendas of strategic issues, challenges, and aspirations, unhitched to any limiting vision. Cohesive teams are needed for day-to-day issues, but learning groups of managers are necessary for spontaneous, self-organizing networks that have open, public conflict and dialogue that are vital to handling the organization's strategic issues. The self-organizing political network of contacts often undermines the hierarchy or bureaucracy. In the absence of this tension and paradox, there can be no change.

Normal management relies on logical analysis, but extraordinary management is exploratory, experimental, and reasons by analogy. Benchmarking against milestones and taking corrective action means business planning by rules, systems, and rational argument. The control and development of an open-ended, unknowable long-term is a political process where constraints are provided by the need to build support; thus, it is self-policing where new strategic directions emerge from the chaos of challenge and contradiction. Top management does not merely drive the process, for it creates favorable conditions for complex learning and effective politics. Master planners create new models and maps by allowing people to learn, and strength grows from contact with the environment, not only from existing strengths.

BUSINESS PLAN SOFTWARE: PROS AND CONS

When creating a Business Plan, there are many software products available, both to install or use over a website, which can assist in the process. The many business plan software products can assist many entrepreneurs unfamiliar with Business Plan format to create their plan more quickly. However, you need to carefully weigh the pros and cons of using business plan software for your own unique situation.

- **The Pros:** A flexible and well-designed business plan software program can take the raw material of your ideas and numbers and arrange these into a standard business plan template. This can prove to be an incredible time-saver, especially to create pro format financial statements for the company. Your knowledge of the accounting that goes into financial statements can be very limited in this case, and the software can walk and talk you through what you need to decide and then generalize and generate the statements automatically. A good software program should offer some ability to customize the look of the plan and to use terms specific to the type of company, product, or service, consumer or business sales, and so forth. Also, the software should allow you to edit the narrative to improve the wording if it is generated automatically.

- **The Cons:** In the final analysis, the integrity of the narrative of the plan and the logic of the ideas within rely on you and your planning team. As the old adage aptly states, "garbage in, garbage out." If you are uncertain about the feasibility of your basic concept, along with how to research the market opportunity or where to find your start-up costs, you need expert help from a business planning consultant or find a partner with more business acumen. If you are unsure as to what you are doing with the organization and how to grow the business, then business planning software will only help you write a plan faster, but can't make your launch successful or improve your chances of raising funds. Furthermore, the ability of any software program to customize your plan will be limited to a certain extent. Software may also offer a basic version and a premium version in an attempt to lure users at a low price and then upgrade them when they realize they cannot make basic changes to the look or content of the plan. You will need to watch out for these types of arrangements and evaluate the success that *real* clients have had finding funding with plans produced by the software, in order to know if it is worthy of purchase.

OUTLINE OF A TYPICAL BUSINESS PLAN

The following is a guideline to the headings that make up a typical Business Plan:

- **Cover Letter:** The cover letter includes basic information about the organization—the address, date of the document, who approved it, who prepared it, and so on.
- **Executive Summary**
    - Organization's purpose and objectives
    - Products and/or services description
    - Competitive advantages of the products and/or services the organization will be providing
    - Present status of the product and/or services
    - Present management team backgrounds and why they bring value to the organization
    - Key success factors
    - Income projections by year

    - Financial investment to date
    - Present financial investment of key people in the organization
    - Financial status and future requirements
    - Present financial needs
    - Profit and Loss Projection
    - Cash Flow Projection
    - Key milestones
    - Assumptions and comments
- **Two-Year Market Analysis**
    - Target market
    - Market research
    - Market trends
    - Price impact on the market
    - Percent of the market the organization can get by year and why
    - Customer profile
    - Marketing and sales strategy and plans
    - Marketing and sales plan strengths and weaknesses
    - Public relations plan
- **Competitive Analysis**
    - Current and future state analysis of the product
    - Competition description
    - Competitive products/services
    - Competitive research data
    - Competitors' strengths and weaknesses
    - Projected changes in the major competitors' product
    - Comparison of organization's product and/or services to those provided by the competition
    - Risk analysis
- **Products and/or Services**
    - Product/service description
        - Positioning of products and/or services
    - Development status of products and/or services
    - Project plan
    - Work breakdown structure
    - Cost to provide the product and/or services
        - Competitive evaluation of products and/or services
    - Delivery schedule
    - Implementing strategy
    - Plans for additional products and/or services

- **Marketing and Sales**
  - Marketing strategy
  - Sales strategy
  - Advertising approach
  - Announcement plans
- **Operations**
  - Mission and visions
  - Organizational structure
  - Business model
  - Intellectual capital
  - Legal structure
  - Insurance
    - Human resources plan
  - Financial management strategy
  - Information technology plan
  - Customer support plan
    - Facilities today and future requirements
  - Organizational SWOT analysis

As you can see, preparing the start-up Business Plan requires a great deal of thinking and work on the part of the key people in the organization, but the time devoted to it is well spent for it forces the leaders of the organization to get organized. Also, the work done on it will serve as a major input to the Strategic Business Plan, which will be prepared as the organization matures.

> The Business Plan is the starting point for a new organization. Do it first as it will make you think it through.
>
> **—H. James Harrington**

7

The Strategic Business Planning Process

> We are all part of the future. Shouldn't we all be part of the planning for it?
>
> **—H. James Harrington**

Strategic management, another term for *enterprise strategy* and *strategic planning* is the art, science, and craft of formulating, implementing, and evaluating cross-functional decisions that will enable an organization to achieve its objectives.* It is the process of specifying the organization's mission, vision, and objectives; developing policies and plans, often in terms of projects and programs, which are designed to achieve these objectives; and then allocating resources to implement the policies and plans, projects and programs. Strategic management seeks to coordinate and integrate the activities of the various functional areas of an enterprise in order to achieve organizational objectives. A *balanced scorecard* is often used to evaluate the overall performance of the enterprise and its progress toward objectives.

> Leadership focuses more on people than things; on the long-term rather than the short term; on values and principles rather than activities; on mission, purpose and direction rather than on methods, techniques and speed.
>
> **—Stephen Covey, Principle-Centered Leadership**

Strategic management is the highest level of managerial activity. Strategies are typically planned, crafted, or guided by the CEO, approved or authorized by the board of directors, and then implemented under the supervision of the organization's executive management or senior executives. In the

* F. David, *Strategic Management* (Columbus, OH: Merrill Publishing Company, 1989).

field of enterprise administration, it is useful to talk about *strategic alignment* between the organization and its environment or *strategic consistency*.*

As strategy today is a subject of application, rather than a discipline, the obvious underpinning disciplines for strategy are economics and organizational sociology. Organizations need to employ them to define a structure in which the process of strategy formulation and its implementation are bound together. Most people are so busy working for their organizations, or are so knee-deep in their organization, that they never find time to work on their organization. Thus they fail to anticipate what might happen or what they might be able to make happen. Unless you regularly schedule time (a one-day, out-of-the-office meeting a month, at least) to work with the Strategic Business Plan and answer critical questions, you'll never achieve your stretch goals.

To be successful today, your organization must become competitor oriented. You must pursue the right competitive strategy, avoid the strengths of your competitors, and look for weak points in their positions. Then you must launch marketing attacks against their weak points in order to achieve and maintain competitive advantage. Sustainable competitive advantage † is the prolonged benefit of implementing some unique value-creating strategy based on a unique combination of internal organizational resources and capabilities that cannot be replicated easily or at all by competitors.

A goal without a plan is just a wish.

—Antoine de Saint-Exupéry

APPROACH TO DEVELOPING A STRATEGIC BUSINESS PLAN

The Strategic Plan provides a roadmap for how the following four questions will be addressed:

1. What do we do?
2. How can we beat or avoid competition?

* See http://en.wikipedia.org/wiki/Business_strategy for more details.

† See the Enterprise Management coach at http://www.1000advices.com/index.html. Their slogan: Free smart & fast advices and inspirational tips from great achievers.

3. How do we excel?
4. What kind of culture do we want?

It typically applies to a 5-year period, but often it is extended to 10–20 years. The Strategic Plan defines where the organization wants to go and how it will get there.

We suggest using a 10-activity approach to develop a Strategic Business Plan. These activities are

- Activity 1: Define the strategy scope and time frames.
- Activity 2: Define the assumptions
- Activity 3: Review the mission, long-term vision, and value statements.
- Activity 4: Define the short-term vision.
- Activity 5: Define core competencies and capabilities.
- Activity 6: Perform risk analysis.
- Activity 7: Define critical success factors.
- Activity 8: Set objectives and goals.
- Activity 9: Develop strategies.
- Activity 10: Develop tactics for each strategy.

One of the decisions that needs to be made early in the strategic planning cycle is whether the organization will be a product-centric or customer-centric organization; this will be one of the primary drivers of the planning and implementation process. Some of the key differences are shown in Figure 7.1.

Product-Centric Organization	Customer-Centric Organization
New products	Customized products and services
Leading-edge/new features	Customized solutions
Sophisticated customers	Preferred/profitable customers
Best products	Best solutions
Market-driven pricing	Value-driven pricing
Understand the competition	Understand the customer
Product teams	Customer teams
Product profit centers	Customer segments
New product development focus	Customer relationship management focus
Asks: How many ways can the product be used?	Asks: What combination of products is best?

FIGURE 7.1
Product-centric versus customer-centric organizations.

- **Product-centric organization:** *Product-centric organizations* are organizations that have multiple product lines that are divided into separate business lines and/or models. Often there are few or no interrelationships among the product lines; for example, a computer manufacturer that sells management consultant services also. They are driven by product portfolios.
- **Customer-centric organization:** *Customer-centric organizations* have offerings that combine and integrate advice, services, and/or software in support of their products and services. They are driven by customer portfolios.
- **Customer-focused organization:** *Customer-focused organizations* use extensive market research in the defining their product offerings and design. They build a demand for the products they are able to produce. They invest in providing their frontline employees with the knowledge, products, and tools to provide effective and consistent customer service.

> What keeps companies from becoming as customer-centric as they should be, want to be, or claim to be? In a word, it's nobody's job.
>
> —**Curtis N. Bingham, *Handbook of Business Strategy 2005***

Activity 1: Define the Strategy Scope and Time Frames

The time period for the most useful Strategic Business Plans is a 5-year to 10-year window. Things are changing so quickly that anything longer is not feasible or realistic. We like to use a 5-year strategic business planning cycle that is evaluated and expanded each year to provide more detail for the remaining years of the plan. Remember, the result of a well-developed and executed Strategic Business Plan is to develop a competitive advantage. Just what is a competitive advantage? Simply stated, it is the answer to the question: What can your organization potentially do better than any other?

Understanding your competitive advantage and the time frames involved are critical. It is the reason you are in business. It is what you do best that draws customers to buy your products and services instead of your competitor's. Successful organizations deliberately make choices to be unique and different in activities that they are really, really good at, and they focus the lion's share of their energy in these areas. You may decide to incorporate your competitive advantage into your mission and/or vision statements.

> What business strategy is all about; what distinguishes it from all other kinds of business planning is, in a word, competitive advantage. Without competitors there would be no need for strategy, for the sole purpose of strategic planning is to enable the company to gain, as effectively as possible, a sustainable edge over its competitors.
>
> **—Keniche Ohnae, *Poor Man's College***

The strategy is about where the enterprise is trying to go in the long term and the planning scope deals with the following questions:

- Which markets should an organization compete in and what kind of activities are involved in such markets? Market scope is of prime importance.
- How can the organization perform better than the competition in those markets?
- What resources (skills, assets, finance, relationships, technical competence, facilities) are required to be able to compete?
- What external, environmental factors affect the organization's ability to compete?
- What are the values and expectations of those who have power in the organization?

Strategies exist at several levels in any organization, ranging from the overall organization (or group of businesses) to the individuals working in it. Corporate strategy is concerned with the overall purpose and scope of the organization to meet stakeholder expectations. This is a crucial level since it is heavily influenced by investors in the organization and acts to guide strategic decision making throughout the business. Corporate products and services are stated explicitly in a *mission statement.*

Strategy business planning is concerned more with how an organization competes successfully in a particular market. It concerns strategic decisions about choice of products, meeting needs of customers, gaining advantage over competitors, exploiting or creating new opportunities, and so on. Finally, the Strategic Plan and organizational alignment are concerned with how each part of the organization is organized to deliver the organization and organization's unit-level strategic direction. Strategic business planning, therefore, focuses on issues of resources, processes, people, and so on.

We like to develop a Strategic Plan every five years. These plans are reviewed annually to be sure they are still relevant.

> Bite off more than you can chew, then chew it. Plan more that you can do, then do it.
>
> **—Anonymous**

Planning Tool 1: Organizational Planning Analysis

We find the following analysis format is very helpful in setting the stage for the organizational alignment cycle.

Instructions: Please read each statement and indicate the extent to which it describes the reality in your organization as a whole. Your responses should reflect what you have experienced as well as what you have generally observed.

Evaluate each statement using a 10-point scale. The left side of the scale indicates that you *totally disagree* and the right side that you *strongly agree.* If you do not know, indicate so next to the question.

1. Your organization has a Strategic Plan that is up to date.
2. Everyone in the organization knows the key points in the Strategic Plan.
3. The Strategic Plan is based upon a list of assumptions that reflects the changes in the environment, customers, competition, and technologies over the coming 5 to 10 years.
4. The organization has well-defined organizational objectives for the next 5 to 10 years and these objectives are communicated to the employees.
5. The Strategic Plan is funded.
6. The individual work-unit measurements are based upon the Strategic Plan.
7. The individual work-unit budgets are related to the Strategic Plan.
8. There is an implementation plan to support the Strategic Plan.

9. The impact of the last Strategic Plan is measured and evaluated.
10. The current Strategic Plan corrects the problem that the last plan missed.

Totals: Add up the scores for each of the 10 questions: ______

Note: A grand total of 50 or less indicates very serious problems and a total of 80 to 100 indicates no serious issues. A total of 50–80 indicates a need for improvement.

Activity 2: Define the Assumptions

A mission statement and its underlying assumptions form a statement about the organization's purpose. It is useful for putting the spotlight on the assumptions underlying the activities that the organization is presently involved in, and the customer needs it is presently endeavoring to serve. It also serves as a guide for day-to-day operations and as the foundation for future decision making. To define the mission statement assumptions, answer these questions:

- What is our core purpose?
- What are we trying to accomplish for our customers?
- What is our organization's reason for existing?
- What gives us the right to perform in this arena?
- What will our competitors do to capture our customers?
- What outside trends will impact our business?
- How will the technology change over the coming 5–10 years?
- What laws may be changed or issued that will impact our products and our customers?

This is where the work you did in the trend analysis that was defined in Chapter 3 will come in handy.

The most important use of assumptions is to help define a vision of where the organization would like to be at some time in the future (five years) and to define roadblocks and opportunities. They also are used to define what actions are needed to overcome negative trends or to take advantage of any positive trends. For this reason, all the relevant assumptions should be well defined and based on sound data and judgment. All assumptions should be well documented. Any Strategic Plan based on false assumptions will usually fail. Based on our experience, not doing a good job of defining an organization's assumptions is the second biggest cause of strategic planning failure. The first biggest cause is not using the Strategic Plan to drive the Annual Operating Plan.

Strategizing is much more than just visioning, forecasting, and planning. In the aligned organization, all substantive issues and assumptions concerning strategy have been redefined as issues of implementation. Today, strategizing is concerned with the assumptions underlying the internal capabilities of the organization and its external environment. The future-focused subject of enterprise strategy is a set of analytic techniques and assumptions for understanding better, and so influencing, an enterprise's position in its actual and potential marketplace.

The Strategic Business Plan needs to be based upon a group of defined, documented assumptions (projected future trends). Taking the time to project future trends is key to having a good Strategic Plan. Some organizations consider the assumptions from three points of view:

- Worst case
- Most probable
- Best case

They then analyze each condition based upon each of the three points of view before they prepare their action plans. For high-risk consideration, worst-case mitigation plans are also developed. Typical future trends that an organization's assumptions can be based upon are

- Demographics
- Lifestyle
- Spirituality
- Financial services
- Internet
- Social responsibility
- The workplace
- Biotechnology and health
- Learning and education
- Products and technology
- Environment
- Politics and government

- Quality
- Associations
- Competition
- Resources
- Transportation

A typical example of the set of assumptions that would affect a non-profit organization is that developed by the American Society for Quality in their Future of Quality Study. They were

- Impact of Globalization
- Increased Social Responsibility
- New Dimensions for Quality
- Impact of Aging Population
- Changes in Healthcare Demand
- Environment concerns
- 21st-Century Technology
- Velocity of Change
- Changing Values
- Information Revolution
- Knowledge Focus
- Shifting Demographics
- Impact of Outsourcing

Assumptions are very important as they are the foundation on which the Strategic Plan is based. Be sure that they are well documented so that they can be reviewed every year to ensure that they were correct or that the outside conditions have not changed, making the assumptions obsolete.

> A strategic conversation with your customers will help you understand opportunities or threats that you have missed.
>
> **—Michael Oleksak, *Handbook of Business Strategy 2005***

Activity 3: Review the Mission, Vision, and Value Statements

We used the word *review* because we believe most of the organizations using this book will already have well-established documented mission, visions, and value statements. If this is not your case, you have a lot more work to do, but the following explanations will provide you with the information you will need to start from scratch and develop your own set of statements.

A review of your mission, long-term vision, and value statements is conducted to be sure they still reflect the direction in which the organization is heading and to be sure that the people doing the planning truly understand not only what they say but also the meaning behind the words.

Reviewing the Mission Statement

While the existence of a shared mission is extremely useful, some strategy specialists question the requirement for a written mission statement. However, there are many models of strategic planning that start with mission statements, and we believe a mission statement is a must for every organization. In fact, a mission statement is a fundamental document that should have been created when the organization was formed.

Many people mistake a vision statement for a mission statement and it becomes a *mis-vision*. The vision describes a desired future state while the mission describes the core purpose and serves as an ongoing guide. A mission statement defines the purpose (type of product or service) or broader goal for being in existence or in the business and usually remains the same for decades, if crafted well. A vision statement is more specific in terms of both the future state and the time frame. Vision describes what will be achieved if the organization is successful.

A mission statement resembles a vision statement in a few companies, but that can be a grave mistake. It can confuse people. The vision statement can galvanize the people to achieve defined objectives, even if they are stretch objectives, provided the vision is a SMART* mission. A mission statement defines what the organization does and does not do. These statements have a direct bearing on the bottom line and success of the organization over a specific period of time.

A mission statement does not need to be changed unless there is a major change in the product or outputs that the organization provides. For example, if an organization produces toothpaste, its mission statement could be: "To provide economically priced products to clean teeth and sweeten breath." If the Strategic Plan called for developing and selling an electric toothbrush, the new product would still be in line with the mission statement. If they decided to produce a mouthwash, it still would be

* SMART is an acronym for specific, measurable, achievable, relevant, and time-bound.

in line with the mission statement. But if they decided to produce deodorant, then it would not be in line with the mission statement and the product should be eliminated from the Strategic Plan or the mission statement should be changed.

The following is Boston Scientific's healthcare mission statement:

> Boston Scientific mission is to improve the quality of patient care and the productivity of health care delivery through the development and advocacy of less-invasive medical devices and procedures. This is accomplished through the continuing refinement of existing products and procedures and the investigation and development of new technologies that can reduce risk, trauma, cost, procedure time and the need for aftercare.

We believe this mission statement is too long; it should have stopped after the first sentence. The mission statement need not include how things will be accomplished.

IBM's mission statement is a good example of a comprehensive statement.

> At IBM, we strive to lead in the invention, development and manufacture of the industry's most advanced information technologies, including computer systems, software, storage systems and microelectronics. We translate these advanced technologies into value for our customers through our professional solutions, services and consulting businesses worldwide.

We believe this could have been shortened to read: "IBM's mission is to service the information handling needs of organizations around the world."

Review the Long-Range Vision Statement

The organization develops a vision statement to paint a picture of how it wants to position itself 10 to 25 years in the future. Top management usually prepares the organization's vision statement. It primarily focuses on the organization's output and how its customers and the rest of the world perceive the organization. The vision statement must be able to express an energizing picture of the future in terms of market presence and customer interface, and it must have enough realism to it to make it aggressively believable.

Jack Welch, former CEO of General Electric stated, "Leaders—and take everyone from Roosevelt to Churchill to Reagan—inspire people with clear visions of how things can be done better. Some managers, on the other hand, muddle things with pointless complexity and detail. They acquaint it with sophistication, with sounding smarter than anyone else. They inspire no one."

> The very essence of leadership is that you have to have a vision. It's got to be a vision you articulate clearly and forcefully on every occasion. You cannot blow an uncertain trumpet.
>
> **—Father Theodore Hesburgh, former president of the University of Notre Dame**

Too often organizations prepare unrealistic vision statements using phrases such as *the best*, *the largest*, and *the most profitable*. Each of these phrases can apply to only one organization in the organization's field of competency. Much better are phrases such as *world class*, *recognized as a leader*, *one of the best*.

Organizations that excel set high standards for themselves and their employees. The following are some best practice guidelines for a hypothetical organization called Excel, Inc.

The following was developed by Abdul Rahman Awl, senior vice president of the Harrington Institute, based upon a book by Claus Moller, a Danish professor.*

- Mission Statement
  - Excel Inc. has a well-defined mission statement, a clear Business Plan, and a long-term strategy for existing.
  - The mission statement is known by everyone.
  - Excel Inc. deploys resources only in areas that naturally fit its mission statement.
- Vision
  - Excel Inc. has a vision: a major, valuable, ambitious goal clearly described as a desirable state achieved at a particular moment.
  - Management inspires commitment to a clear vision and definite objectives.
  - The vision is communicated to everyone in such a way that all know it and understand it.
  - The vision is sufficiently ambitious and meaningful to enable everyone to contribute to its implementation.
  - The vision penetrates the life of the organization, mobilizes energy, and creates a passion for success.

---

* Claus Moller, *Energy Meter* (Hillerod, Denmark: TMI) 1995.

President John F. Kennedy gave a good example of a convincing and challenging vision when he declared that the United States would send a man to the moon and bring him back safely before the end of the 1960s.

Another good example is CNN's long-term vision statement: "To provide firsthand news from all around the world as it is happening, customized to each country's interest and needs."

Review the Value Statement

The short- and long-term vision statements must be in line with the value statement of the organization. If they are not, one of the two needs to be changed. In most cases it is easier to change the vision statements than the value statements because the value statements drive the culture of the organization and it takes a long time to change culture.

- **Values:** These are things that an organization holds as sacred—things that should never be sacrificed. Much care should be taken when an organization selects its value statements to be sure they do not conflict with each other. Values are often interchanged with beliefs, but management is held accountable for complying with values.
- **Principles:** Principles are the ultimate source, a fundamental truth, the motivating source upon which others are based. If a statement is ever not true or not followed, it is not a principle. (Note: Any manager or employee who is not living up to the organization's principles, is a good candidate for separation from the organization.)
- **Beliefs:** A belief is an acceptance of something as being true. It, in fact, may or may not be true.

There is a great deal of confusion about the use of these three terms (principles, values, and beliefs.) I believe that you need to be very careful about the terms you use. Do not use the term *principles* unless you and the rest of the management team agree that they are statements that they would never compromise. This is the reason that we like the term *values*, which allows for using some judgment. We feel that the term *belief* is a weak statement. I believe that it is good to lose weight, but I do nothing to make it happen. (Anyone hungry for a hot fudge sundae?) It is critical that everyone in the organization has an understanding of the foundation on which the organization's culture is based. These statements provide direction to management that governs their performance and their behaviors.

To the employee, they provide a promise of conditions on which the organization is built.

Let's look at Ford Motor Company's basic values:

- **People:** Our people are the source of our strength. They provide our corporate intelligence and determine our reputation and vitality. Involvement and teamwork are our core human values.
- **Products:** Our products are the end result of our efforts and they should be the best in serving customers worldwide. As our products are viewed, so are we viewed.
- **Profits:** Profits are the ultimate measure of how efficiently we provide customers with the best products for their needs. Profits are required to survive and grow.

Ford's guiding principles are

- **Quality comes first:** To achieve customer satisfaction, the quality of our products and services must be our number one priority.
- **Customers are the focus of everything we do:** Our work must be done with our customers in mind, providing better products and services than our competition.
- **Continuous improvement is essential to our success:** We must strive for excellence in everything we do: in our products, in their safety and value—and in our services, our human relations, our other competitiveness, and our profitability.
- **Employee involvement is our way of life:** We are a team. We must treat each other with trust and respect.
- **Dealers and suppliers are our partners:** The Company must maintain mutually beneficial relationships with dealers, suppliers, and our other business associates.
- **Integrity is never compromised:** The conduct of our Company worldwide must be pursued in a manner that is socially responsible and commands respect for its integrity and for its positive contributions to society. Our doors are open to men and women alike without discrimination and without regard to ethnic origin or personal beliefs.

Note: Ford developed two sets of statements—one was "defined or basic values" and the other set of statements was called "guiding principles."

The following is a quote from IBM's Managers Manual:

> An organization, like an individual, must be built on a bedrock of sound beliefs if it is to survive and succeed. It must stand by these beliefs in conducting its business. Every manager must live by these beliefs in making decisions and in taking action.

The beliefs that guide IBM's activities are expressed in IBM's principles.

- **Respect for the individual:** Our basic belief is respect for the individual, for each person's rights and dignity.
- **Service to our customer:** We are dedicated to giving our customers the best possible service. Our products and services bring profits only to the degree that they serve the customer and satisfy customer needs.
- **Excellence must be a way of life:** We want IBM to be known for its excellence. Therefore, we believe that every task, in every part of the business, should be performed in a superior manner and to the best of our ability. Nothing should be left to chance in our pursuit of excellence.
- **Managers must lead effectively:** Our success depends on intelligent and aggressive management, which is sensitive to the need for making an enthusiastic partner of every individual in the organization.
- **Obligations to stockholders:** IBM has obligations to its stockholders whose capital has created our jobs.
- **Fair deal for the supplier:** We want to deal fairly and impartially with suppliers of goods and services.
- **IBM should be a good corporate citizen:** We accept our responsibilities as a corporate citizen in community, national and world affairs; we serve our interests best when we serve the public interest. We believe that the immediate and long-term public interest is best served by a system of competing enterprises.

The basic principles for the Naval Ship Weapon Systems Engineering Station (NSWSES) are called their *operating principles*. These principles, approved by Captain J. J. Andrilla, US Navy Commanding Officer, are as follows:

- **People:** We respect each person's rights and dignity. Teamwork and cooperation are the way we get the job done. We reward employee achievements. We are committed to helping employees reach their full potential. Differences in employee abilities and background are a source of our strength.

- **Quality:** We are committed to never ending improvement. We believe in total quality management. Quality improvements lead to increased productivity, reduced costs and overall enhanced performance. Consistent quality products = customer satisfaction = our reputation.
- **Customers:** We are dedicated to giving our customers, internal and external, the best possible service. Our outputs bring true benefit only to the degree that they serve the customer and satisfy customer needs. Our existence depends upon their satisfaction. Success is when our customers brag about us.
- **Excellence:** We want excellence as a way of life at NSWSES. We believe that every task, in every part of our mission, should be performed in a superior manner and to the best of our abilities. Nothing should be left to chance in our pursuit of excellence.
- **Leadership:** Our success depends on intelligent and aggressive management which is sensitive to the need for making an enthusiastic partner of every individual in the organization. We must plan for the future by keeping an open mind to new ideas, whatever the source.
- **Suppliers:** We want to deal fairly and impartially with suppliers of products and services. Selection of suppliers must be according to the quality of their products or services, their reliability and competitiveness of price.
- **Workplace:** We are committed to improve our facilities and tools for the betterment of our people, processes, products and safety.
- **Our ethic:** We are honest and straight forward in all our business dealings. We have an obligation to the U.S. public whose taxes have created our jobs. We deal with each other, our customers, and our community with integrity and pride.

The US government has had a set of basic principles since the beginning. It is called the *Bill of Rights* and it has helped guide the United States for more than two hundred years.

These fundamental statements that we will call basic principles are the things that attract new employees to your organization. They define the rules that the organization will not compromise. They are the employees' *bill of rights* so they should be changed infrequently, and then only when they have become obsolete because of the social or external environment. Dr. Harrington worked for IBM for forty years and their basic principles remained constant throughout that time period.

Texaco defines its values as

- Establish the self-image that we can be the best.
- Require profit accountability of all managers.
- Set up objectives that affect the good of the organization.
- Instill a sense of urgency to do it today.
- Know where the problem is and solve it.
- Challenge how we have always done it.
- Develop heroes and reward success.

We recommend that an organization define 8 to 12 value statements.

Management's job is to live every business minute complying with the organization's basic principles. This is the first obligation of every manager. No employee should accept a management position unless they have already lived up to these basic principles, and no manager should be left in a management role unless they live up to these basic principles. Employees have a responsibility to work in support of the basic principles, refusing to compromise them. Managers who ask their employees to bend the basic principles are not doing their job, and this condition should be brought to the attention of upper management.

Activity 4: Define the Short-Term Vision

A short-term strategic vision is the image of an organization's future—the direction it is headed, the customer focus it should have, the market position it should try to occupy, the organizational activities to be pursued, and the capabilities it plans to develop. Forming a short-term strategic vision should delineate what kind of enterprise the organization is trying to become at the end of the strategic time frame 5 to 10 years into the future. This will infuse the organization with a sense of purposeful action. Think big! To write a short-term vision statement, answer these questions: What are the key factors underlying our current situation and what should our organization look like 5 years from now? What are the assumptions that the plan is based upon? What is the desired future state in both the short term and the long term?

It is useful to begin the review with a SWOT (strengths, weaknesses, opportunities, threats) analysis, which helps you look critically at your organization's mission, long-term vision, and values. It is a tool to help produce a good fit between an organization's strengths and its opportunities. Assess your strengths and weaknesses by answering these questions:

- What do we do best?
- What do we not do best?
- What are our organization's resources—assets, intellectual property, and people?
- What are our organization's capabilities (functions)?

Assess your opportunities and threats by answering these questions:

- What is happening externally that will affect our enterprise?
- What are the strengths and weaknesses of each competitor?
- What are the driving forces behind sales trends?
- What are important and potentially important markets?
- What is happening in the world that might affect us?

If you have an established organization where the mission has been defined, then the mission guides the vision statement and the rest of the Strategic Business Plan. In either case, leaders need to know the fundamental purpose: the mission, current situation in terms of internal resources and capabilities (strengths and weaknesses) and external conditions (opportunities and threats), and where they want the enterprise to go—the vision for the future.

Some features of an effective vision statement include

- Clarity and lack of ambiguity
- Vivid and clear picture
- Description of a bright future
- Memorable and engaging wording
- Realistic aspirations
- Alignment with organizational values and culture

To become truly effective, an organization's vision statement must become aligned with the organization's culture. Leaders have the responsibility to actively communicate the vision regularly, create narratives that illustrate the vision, act as role models by embodying the vision, create short-term objectives compatible with the vision, and encourage others to craft their own personal vision compatible with the organization's overall vision.

Many organizations do not have a well-understood vision around which their Strategic Plan is designed. They are a lot like a man driving a car with a muddy windshield who can't see the road. He doesn't know whether to turn right or left or to slow down or speed up. A good vision statement is like cleaning the mud off the windshield.

The following was Toyota's global vision for 2010, which it called "Innovation into the Future—A Passion to Create a Better Society."* This vision was announced in April 2002.

- Through "*monozukuri*—manufacturing of value-added products" and "technological innovation," Toyota is aiming to help create a more prosperous society. To realize this, the company is challenging themes:
  - Be a driving force in global regeneration by implementing the most advanced environmental technologies.
  - Create automobiles and a motorized society in which people can live safely, securely, and comfortable.
  - Promote the appeal of cars throughout the world and realize a large increase in the number of Toyota fans.
  - Be a truly global company that is trusted and respected by all peoples around the world.

CNN's vision is a long-term mission statement and Toyota's is a short-term vision statement.

The purpose of this activity is to compare the outputs generated so far to ensure they are in line with the general product and service guidelines defined in the mission statement. The next part of this activity is to evaluate if the short-term vision statement is aggressive enough to support the long-term vision statement, and lastly, to ensure the plan is in tune with the organization's values.

Activity 5: Define Core Competencies and Capabilities

All successful organizations have the ability to identify four key areas of strategic focus that are characterized by the following:

* Toyoland.com

- Technologies or functions of the organization that define its specific areas of excellence
- Activities that are most valued by their customers
- Areas in which the organization is especially good
- Areas that will distinguish the organization from its competition

Through the years, these strategic focus factors have been given under many names. The most common are

- **Core competencies:** The technology and product skills that an organization's products and/or services are based upon (for example, Sony's skill at miniaturization).
- **Core capabilities:** The business processes that visibly provide value to customers (for example, Honda's dealer management process)
- **Strategic excellence positions:** Unique and distinctive capabilities that are valued by the customer and provide a basis for competitive advantage (for example, Avon's distribution system)

Call them what you will, but to excel, the organization must focus its resources on improving these key attributes to ensure that the organization keeps ahead of its competition in these critical areas. The organization can farm out everything else that it does, but it must preserve its leadership role in these strategic focus functions. Typically, the strategic focus is defined by top management and looks five years in the future.

Activity 6: Risk Analysis

A lot of things can go wrong—changes in the organization's environment can occur outside the organization's control. People who prepare the Organizational Master Plan must be like chess players; the best chess players not only plan their moves far in advance, but they also seek moves that offer them the greatest flexibility to take advantage of future unexpected events. The organization must be flexible enough to take advantage of and deal with unanticipated internal and external events. Typical risks that must be considered are

- Competitive initiatives
- Too many initiatives

- Sponsor situations
- Customer needs
- Organization's economics
- Supply problems
- Industry trends
- Stock market fluctuations
- Wars
- Government rulings

Figure 7.2 shows the seven phases that make up the risk management cycle.

Successful risk management ensures that your organization will identify, analyze, and respond to risk. In global organizations, the element of risk can be greater because of team members from one country assuming that things can be done the same way in another country as at home. This puts the onus on the senior manager to validate plans in the country where they will be executed. An experienced global leadership team can minimize these risks as it develops the expertise to trap faulty assumptions early in the strategic planning sessions. Communications play a big role in minimizing internal risk based on poor planning and false assumptions. Good communications also play a big role in understanding and planning the management of external risks over which the team has no control, such as weather and fluctuations in currency exchange rates.

Murphy's Law—anything that can go wrong will go wrong—is not necessarily accepted as widely in some countries as it is in North America or the United Kingdom. The denial of risk's negative outcomes tends to arise in certain cultures in which all decisions are made by top management, and corporate belief in excellence allows little room to admit that errors can and do occur.

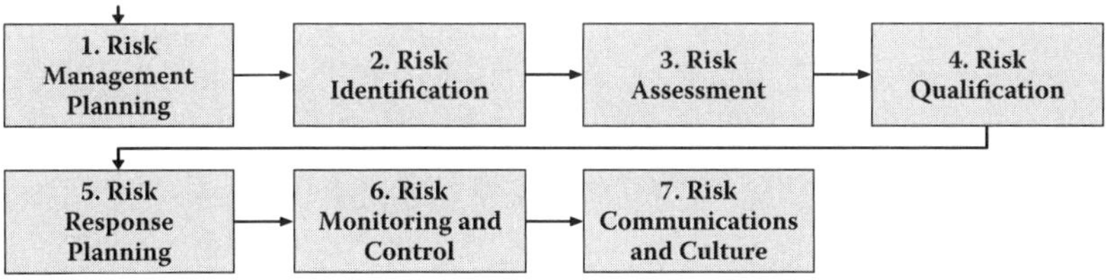

FIGURE 7.2
The seven phases of a risk management cycle.

Activity 7: Critical Success Factors

These are defined by top management and look forward 2 to 3 years in the future. They include the traditional strengths, weaknesses, opportunities, and threats (SWOT) analyses. Preparing a vision statement requires the executive team to look outside the box and take an unconstrained strategic perspective. It's also important that the vision be linked back to today's reality by specifically focusing on the things that could affect or prevent the successful implementation of the plan. The critical success factor defines the conditions on which the plan is based. These can vary all the way from insufficient funding to lack of market growth. Critical success factors perform the service of identifying obstacles that need to be addressed if the plan is to be successfully carried out. They often create several additional strategies that are incorporated into the Strategic Plan. Often, critical success factors are translated into risks for which mitigation plans need to be prepared. By identifying these critical success factors and monitoring them closely, the probability of successfully carrying out the Strategic Plan is greatly improved. The organization needs to view these obstacles as additional challenges, not as obstacles that can't be overcome.

> Make no little plans; they have no magic to stir men's blood and probably will themselves not be realized. Make big plans; aim high in hope and work, remembering that a noble, logical diagram once recorded will not die.
>
> **—Daniel H. Burnham, American architect and urban planner**

Activity 8: Setting Objectives and Goals

This category consists of two parts, organizational objectives and performance goals. The words *objectives* and *goals* seem to have different meanings to different organizations. Some organizations believe that you set objectives and develop goals to support the objectives as measurements along the way. Other organizations look at goals as a higher-level measurement and objectives as the milestones along the way to meet the goals. Let's define both based upon Merriam-Webster's Collegiate Dictionary.

- **Objective:** something toward which effort is directed; an aim or end of action
- **Goal:** the end toward which effort is directed; the terminal point of a race

As you can see, they both are end points, so in reality they can be used interchangeably. The key is the word that defines what kind of goal or objective they are. In this discussion, organizational objectives are the higher-level condition, and performance goals are the milestones along the way that are used to measure progress toward the organizational objectives.

Organizational Objectives

These are prepared by top management and approved by the board of directors. They define what the organization wants to accomplish in the next 5 to 10 years. They provide a big-picture focus that is usually not quantified. Typical organizational objectives would be

- To sustain your profitability, which funds your growth
- To exhibit product leadership across your entire product line to excel in technology, value, and quality
- To grow with the industry
- To be the low-cost producer, the low-cost seller, and the low-cost administrator
- To quantify the impact that all major decisions have upon the organization's stakeholders

As you can see, these types of organizational objectives are not *quantified* but are *quantifiable*. For example, to grow with the industry can be quantified; if the industry grows at a rate of 15 percent in a given year, the organization should grow at least 15 percent during the same year.

It's very important that you communicate the organizational objectives to each and every employee within the organization and that you continuously report back to all employees any progress related to each of these objectives. Good organizational objectives that are not followed and reported are just scraps of paper that are no good to anyone.

In the early 1960s, John F. Kennedy set an objective to put a man on the moon by the end of the decade. He didn't know how it would be done or how much it would cost. He knew there were a lot of problems that needed to be solved—technical, financial, and political. He just set the objective and defined the end date. Then he left it up to others to define intermediate goals and strategies. He stepped back and got out of the way so that others could do their thing, but he never lost interest or stopped following the project. Everyone knows what the result was—before the end of the decade, Neil Armstrong took his "giant leap for mankind."

Objectives set by management define what and when major accomplishments should be achieved. Once that is done, get out of your employees' way so they can get the job done. Don't think that you need to have all the answers in advance of setting the organization's objectives.

In the early 1980s, John Young, then president of Hewlett-Packard, set an objective to improve all of HP's products by 10 times in 10 years. That's a 1,000 percent improvement! Engineers and management were overwhelmed with the task. They told him it couldn't be done the way they were working. John's reply was, "Now you got the idea."

For Jack Welch, the legendary former CEO of GE, *visioning* and strategy management (stretching) is doing the best possible—and then reaching beyond. What Welch calls *stretch* simply means determining performance targets that are achievable, reasonable, and within GE's capabilities. In the organizational alignment context, this means raising sights higher—much higher—toward goals that seem almost beyond reach, goals requiring superhuman effort to achieve. Often enterprise leaders set objectives and exceed goals even as they fall short of the stretch. "Don't punish, reward them," says Welch, "Once you have the right 'themes' in place, that is." What is critical involves the setting of the enterprise performance management bar high enough; otherwise, it's impossible to find out what people can do.

One of the organizations we worked with developed eleven organizational objectives. These included

- Gaining employee commitment, winning their support, loyalty and participation through the application of best human resources practices to build superior performance within every function and level.
- To develop a dynamic reconfigurable organization that recognizes and responds rapidly to the changing needs of Dubai.

Organizational objectives can be general or focused. Examples of stretch objectives are

- John Young, HP's former vice president, set stretch objectives to cut failure rates to one-tenth the current performance in a 10-year period.
- William J. Weisz, Motorola's chief operating officer (COO), set an objective of tenfold improvement in 5 years.
- In 1994, one of Hughes Aircraft divisions set a 10 times improvement objective in one year.

Performance Goals

These have two key elements: first, they specifically state the target for improvement, and second, they give the time interval in which the improvement should be accomplished. They usually have a short- (one-year) and long-range (five-year) targets. (For example, your short-term goal might be to have 80 percent of your customers rate you as "exceeds requirements or better" by the end of this year, and your long-term goal would be to have 70 percent of your customers rate you as "outstanding" within the next five years.) Each year, middle managers should develop a set of short-range performance goals that are reflected in the department's budget. Top management should review and approve these goals to be sure that they support the organizational objectives and are aggressive enough.

Organizational goals cover such major issues as

- Products offered
- Customers targeted
- Company image
- Competition
- Levels of service
- Product, program, and service quality

Dubai Municipality defined over 350 goals that were scheduled over the coming 5 years in order to meet the 11 organizational objectives that they set for themselves.

It is important to realize that the customer often doesn't realize what they will not need or want in the future. Customers' input is critical for short-term planning (one to three years). The customers' input reflects their current needs; even when you ask them to define their future needs, they will have a limited view of how technology will be changing over time. Your team is the expert in your technology, not the customer; that was the big mistake IBM made in the 1980s.

At the beginning of the 1980s, IBM owned the personal computer (PC) and mainframe markets. When John Akers became president of IBM, their market studies indicated that their customers wanted faster mainframes, not PCs. The engineers at IBM projected that the PC market was going to take off because of the improved technologies that could be developed. IBM listened to its customers and not to the engineers; they invested their development budget in developing faster mainframe computers. As a result, by the beginning of the 1990s, IBM recorded the biggest loss in one year in any organization's history up to that date. The customer knows what they need now, but very often they don't know what they will need in the future related to the products and/or services you can provide to them.

Organizational objectives and performance goals are the alignments to your mission and vision. Realistic actions and objectives for each organizational unit are developed from the SWOT analysis and customer profile. Objectives set the agenda and are broad and global in nature. Write 2 to 10 objectives that give action to your mission/vision and will take a few years to achieve. Then develop goals that are in line with and support each objective. Goals should be measurable, quantifiable, and support your objectives. Think about achieving them in one-year time-frame intervals. Effective goals must state how much of what kind of performance is to be accomplished, the deadline for accomplishing the goal, and who will be responsible for accomplishing it. Make sure both your goals and objectives build on your strengths; shore up your weaknesses, capitalize on your opportunities, and recognize your threats.

> No matter how carefully you plan your goals, they will never be more than pipe dreams unless you pursue them with gusto.
>
> **—W. Clement Stone, author**

Activity 9: Developing Strategies

This is the first item in the action part of the Organizational Master Plan. Once clear targets have been identified, a set of strategies must be decided on to further the organization's efforts. *Strategies* are defined here as specific programs, initiatives, and decisions that will require resources to be allocated to them. They can range from the development of strategic alliances to developing and conducting special in-house training for customer service. These operating strategies are usually very specific since the good plans are very clear and focused. Occasionally there may be some strategies that are critical and yet don't get openly shared, particularly since they may be extremely sensitive and widespread knowledge of them would put you at a competitive disadvantage (e.g., mergers, acquisitions, and dispositions of certain business units). Once the strategies are identified, then lead responsibilities should also be assigned in order to clarify the implementation activities and create accountability in the organization.

The strategies document the approach that will be used to meet the performance goals. They are generated by middle management with input from first-line management and approved by top management. Every effort should be made to keep the strategies up to date, without making major changes. Drastic changes in the strategies upset the organization and require a major expenditure of resources in reacting to those changes. Major changes can also result in the termination of projects that are only partially completed or have not become totally effective. It should be apparent that many strategies are generated by many different functions that support the business objectives.

Strategies are defined as the approaches that will be used to meet performance goals. Merriam-Webster's Collegiate Dictionary defines strategy as "the art of devising or employing plans or stratagems toward a goal." They define *what* is going to be done, not *how* it is going to be done. (For example, you will sign alliance partnerships with a software firm that produces resource management software or you will redesign three of your critical processes that have the biggest impact upon cycle time.) Strategies often stay constant for a number of years. In fact, you should make every effort to keep the strategies up to date without making major changes. Dropping one strategy to pursue another strategy in a different direction often prevents the original strategy from meeting its objectives and wastes many resources.

It is important to note that the objectives and performance goals are usually divided into major functions like marketing, sales, development, production, IT, quality, and so on. The Strategic Plan will include, for each of these functions, a breakdown of the strategies (over the five-year period) that they will be using in support of the objectives and performance goals part of the plan. Typically, the first and second years will have a more detailed, comprehensive breakdown of strategies, with year 3, year 4 and year 5 defining future potential strategies.

Each year during the planning cycle, each of the remaining years should be readdressed to add more "meat to the bones" and to reflect the success or lack of success the organization had in implementing last year's Strategic Plan.

Before you start to do any planning, set some measurements that you will use to measure and gauge the effectiveness and success of the activity. Some typical measurements would be

- Customer satisfaction
- Employee satisfaction

- Communication index
- Operating cost
- Better implementation of the Strategic Plan
- Improved output quality
- Increased market share

Some of the performance improvement approaches that might be used on a typical project are as follows:

- Reduce bureaucracy
- Reduce bottlenecks
- One-stop shopping
- Empower people to make decisions at the lowest possible level
- Decentralization
- Outsource noncore processes
- Reduce layers of management
- Not influenced by policies
- Change management

Strategies set specific approaches that lead to achieving your goals and objectives. Basically, these are write-in, action-oriented to-do lists for each goal. A quick way to develop your strategies is to answer this question: What current situational problems and future roadblocks exist that could keep the organization from achieving its objectives and goals? Use the answer to develop action items for each goal. Assign responsibilities and deadlines to ensure implementation. A useful method to get buy-in from your staff is to assign a goal to key respected members of the staff. Ask them to develop the strategies related to the goal and be responsible for making sure each strategy has a plan developed to implement it.

Because each part of the organization develops it own set of strategies, we recommend that at this point the complete set of strategies are brought together in order to identify any overlaps, similar approaches and/or voids. We often find that a number of departments will come up with the same strategy, and if you do not bring them together, it will lead to a great deal of confusion. When there are a number of different organizations planning on using the same strategy, it is usually best to assign one organization to coordinate the strategy for all the organizations. A good example of this is Lean Six Sigma. In many organizations, like manufacturing engineering, areas such as product engineering, sales, production, finance, and quality assurance may be planning on Lean Six Sigma as a strategy to reduce waste. Now you would not want six different approaches to Lean Six Sigma going on in the organization. The answer is to appoint some area to coordinate each of the common strategies.

Defining Actions

The best laid plans will be nothing but wasted effort if they are not implemented. This is the weak part of many organizations' plans. Lots and lots of work goes into preparing the Strategic Business Plan, but it is not effectively carried out in some organizations and the Strategic Business Plan has very little impact upon the way the organization functions. It is easy to plan, to define what someone else should do, but making it happen requires lots of work, sweat, and dedication. The action phase of the Organizational Master Planning process is designed to focus the organization's resources on the expectations that top management has set and to which the board of directors has agreed.

Four planning elements define how the organization's resources will be used to carry out the Strategic Business Plan. Only strategies and tactics are part of the strategic planning process, but all four are at the heart of the Organizational Master Plan. In many organizations, tactics are only addressed as part of the organization's Annual Operating Plan. The four planning elements include

- Strategies
- Tactics
- Budgets
- Performance plans

Activity 10: Develop Tactics for Each Strategy

The dictionary defines *tactics* as the art or skill of employing available means to accomplish an end. In this case, tactics are the detailed plan that defines how the resources will be used to carry out each strategic business strategy. For each tactic, the time required to complete the tactic will be estimated. They are developed jointly by the first-line manager and the employee who will be implementing them. They normally are detailed down to the activity level but not down to the task level; they fall short of a complete work breakdown structure. The tactics are updated at least once a year and change frequently. Tactics become very personal to the first-line manager and the employee because they define a major part of the coming year's endeavors.

A typical list of tactics for a software system would be

- Prepare the project plan and get it approved: 2 months
- Define the present status: 3 months
- Design the system: 3 months
- Write the software programs: 6 months
- Model the system: 2 months
- Install the system: 3 months
- Measure the results and make any changes needed: 6 months

SUMMARY

The Strategic Business Plan is not top management's plan or the president's plan. If you receive a paycheck, it's your plan. Top management may define the overall game plan, but it's up to each of you to do your part to make it work. You may not agree with the plan; you may even question if it will work. But once the plan is put in motion, you have a responsibility to do everything you can do to make it work. Top management cannot do the job by itself; it requires the whole team to work in harmony to accomplish the tasks required to make the plan a success.

Typical outputs from the strategic business planning cycle are as follows:

- Mission statement
- Long-range vision statement
- Short-range vision statement
- Value statements
- List of assumptions on which the plan is based
- List of core competencies and capacities
- Analysis of risk and mediation plans for all the high-risk items
- List of critical success factors
- List of business objectives
- List of goals related to each business objective

- List of strategies required to meet each of the goals by the part of the organization that is responsible for carrying them out
- List of tactics required to implement each strategy broken down into yearly activities

Calvin London, vice president of INO Therapeutics, reported in *Quality Progress* (August 2002) the following savings for an Australian company that used a good Strategic Plan for a two-year period.

- Sales revenue up 25%
- Export sales over budget 30%
- Reject rates down 81%
- Average unit throughput up 37%
- Overtime cost down 10%
- Lost time injuries down 87%

> Don't treat your Strategic Business Plan like most of us did last year's New Year's resolutions.
>
> **—H. James Harrington**

8

Strategic Improvement Plan

> The lack of good long-range improvement plans has cost the US government trillions of dollars and will cost them trillions more over the next four years.
>
> —**H. James Harrington**

INTRODUCTION

Everyone is talking about the need for a cultural change, but we believe that focusing on that issue is the wrong answer to today's problems and does not prepare most organizations in the developed world to survive in the twenty-first century. *Culture* is defined as your background, your history, your heritage, your religion, and your beliefs. Most organizations want to hold on to their culture and in fact, are worried about losing it. America should be proud of its culture. It is a culture rich in imagination, hard work, caring, risk taking, and accomplishment. It is a culture that made the United States the richest, most powerful, most productive nation in the world. Organizations in Asia and Europe also have national cultures that they want to preserve. And let's not forget the organizations in South America and Africa. Our culture is not the problem. It is the personality of today's population that is the problem. We talk about "workaholics" like work is the worst four-letter word in the English language. People work overtime begrudgingly if they are notified 72 hours in advance, and if not, they refuse. It is the personality of today's work force and our children that needs to be changed. *Personality* is defined as an individual or group's impact on other individuals or groups. We need to change the personality of our people before we lose the culture that our forefathers worked so hard to create. It is the personality of our key managers that dictates the personality of the total organization. When a new CEO is appointed, the total organization adapts to his or her personality.

> If the new CEO is a basketball fan, everyone knows last night's basketball scores.
>
> **—H. James Harington**

We cannot go back to what we used to do, for this old world has changed. The amount of information available to the individual doubles every five years. According to Richard Wurman in his book *Information Anxiety* (2001), the weekday edition of the *New York Times* contains more information than the average person was likely to come across in a lifetime in seventeenth-century England.

How do we create a change in the personality of our people? We do that by changing the environment in which they live. Our personality is molded by an ongoing series of environmental impacts to which we are subjected. It starts at birth and ends with our last heartbeat. The biggest impact occurs during the formative years of a child's life—the period before they enter school (a time when more and more of our children are not spending the amount of time they need with their parents, due to two-salary families and the single-parent family environment in which they live). Our children do not understand quality time. They need full-time parents to help them mold their personality, but that subject is outside the scope of this document.

Let's focus on what we can do to influence and change the environment that impacts the personality of today's work force. What the organization needs is to develop a plan that will change the environmental factors that impact the personality and behavioral patterns of the employees, placing special emphasis on the management team. If we sustain a positive change in the personality of the organization for a long enough period of time (about five years), we will change the organization's culture.

We have done a lot of work with executive teams to change their behavioral patterns. The average executive makes between 50 and 80 behavioral errors per week. Correcting these behavioral problems can do more to improve the organization's performance than any Six Sigma program can.

STRATEGIC BUSINESS PLANS VERSUS STRATEGIC IMPROVEMENT PLANS

There is a big difference between a Strategic Business Plan and a Strategic Improvement Plan.

- The *Strategic Business Plan* sets the product and service strategy for the organization—the markets that it hopes to penetrate, the new products that will be introduced, the production, and so on. It is a plan that directs and guides the business as it is viewed by its customers. The Strategic Business Plan is primarily directed at meeting the needs of only two of the stakeholders associated with the organization—the customer and the stockholder. It is a plan that is primarily focused on the external opportunities.
- The *Strategic Improvement Plan* is an internally focused plan that is designed to transform the environment within the organization and to change its personality (behavioral characteristics) to be in line with the Strategic Business Plan. It takes into consideration the needs of all of the organization's stakeholders from an improvement standpoint. The Strategic Improvement Plan defines the transformation in the business personality of the organization. It provides an orderly passage from one state or condition to another. The Strategic Improvement Plan supports the Strategic Business Plan, so the two of them, although different in content and intent, must be kept in close harmony.

WHY DO YOU NEED A STRATEGIC IMPROVEMENT PLAN?

"Why does my organization need to develop Strategic Improvement Plan? I know a lot of problems that we can start working on right now. In fact, we are already working on them." Don't stop working on your problems. You cannot afford to stop putting out the fires that your present organization is fueling, but as long as you continue to do what you have been doing, you will continue to get the results that you have been getting. Unfortunately, your competition probably is not content with their situation and they are changing, so if you don't change, they will improve their competitive position. In fact, the twenty-first century will be dominated by those organizations that improve the most and can change the fastest.

Most organizations, in their hunger to improve their relative performance, have embraced many different improvement tools. It seems like each time someone went to another conference they came back with another improvement tool. Our research shows that more than 1,100 different improvement tools exist, and this data is not complete. Each of these tools works under the right conditions. Many solve the same type of problems. Figure 8.1 lists a some of the different tools for the management support and leadership category.

No organization can afford or effectively utilize all 1,100 improvement tools. Many of the tools overlap in approach and the problems that they solve. Many of them are not applicable or will have little impact upon your organization. Your team is made up of many different individuals (see Figure 8.2). Just as individuals differ, organizations differ in many ways (see Figure 8.3).

Add to this complexity the fact that winning, surviving, and poorly performing organizations have to do a very different set of things to improve. For example, benchmarking is a must-use tool for most winning organizations, but can drive the poorly performing organization into bankruptcy. It becomes readily apparent that there is no one approach to improvement that is correct for all organizations or even for different sites within an organization. One of the major mistakes made by most organizations that are not in the top 10%, is looking at the best organizations and trying to copy them. This approach often leads to disaster.

One of the most important studies ever undertaken related to performance improvement was conducted by the American Society for Quality Control (ASQC) and Ernest & Young in 1992 and was called the International Quality Study Best Practices Report (published by America Quality Foundation and Ernst & Young in 1993). It is the only study that we know of where sufficient data was collected to statistically evaluate the true effectiveness of the different improvement approaches on return on investment (ROI), value-added per employee, customer satisfaction, and profits. It proved that many of the improvement approaches, which were effective for the top 10% of the organizations, were not beneficial to the other 90% of the organizations, and in many cases, they were detrimental to the poor-performing organizations.

Management Support and Leadership Tools	
1. Management Self-Audits	30. Teams—Group Process
2. Career Planning	31. Leadership Skills Development
3. New Performance Standards	Communication
4. Quality Policy	Listening
5. Quality Company Policies	Inspiring
6. Job Descriptions	Coaching
7. Quality Visions	Modeling
8. Annual Strategic Quality Plans	32. Self-Managed Work Teams
9. Pay for Performance	33. Facilitator Training
10. Performance Planning and Evaluation	34. Flattening the Organization
11. Career Development	35. Responsibility Charting
12. Audits by Top Management	36. 7-S Model (Organizational Effectiveness)
13. Employee Surveys	37. Risk/Opportunity Management Process
14. Gain Sharing	37. Goal Control Process
15. Management by Objectives	39. Personality Profile (Keirsey-Bates)
16. Effective Listening	40. Management's Seven Tools
17. Long-Range Quality Planning	Affinity Diagram (K-J)
18. Poor-Quality Cost	Interrelationship Diagram
19. Nonverbal Communications	Tree Diagram
20. Risk Taking	Matrix Charts
21. Time Management	Matrix Data Analysis
22. Quality Communications	Process Decision Program Charts
23. Quality Policy Deployment	Arrow Diagram
24. Leadership Skills Training	41. Structured Customer Surveys
25. Effective Delegation	42. Quality Area Improvement
26. Meeting Management	43. Line Management
27. Management by Walking Around	44. Cross-Functional Management
28. Participative Management	45. Hoshin Planning
29. Six Sigma	

FIGURE 8.1
Sample of the 1,100 plus improvement tools.

WHAT CREATES YOUR ORGANIZATION'S CULTURE?

The organization's culture is created over a long period of time as a result of the way management implements the organization's basic principles and values, combined with the way the employees react to management's stimulation. Let's look at some definitions before we go further.

EVERYONE IS DIFFERENT

- Some are big, others are small.
- Some are thin, others are tall.
- Some can run, others must crawl.
- Some are strong, others are weak.
- Some are loud, others are meek.
- Some are men and some are boys.
- And some still like to play with toys.
- Some are women, and others are girls.
- Some are bald, and some have curls.
- While some struggle, others are blessed.
- There's just no one right way to dress.

FIGURE 8.2
How individuals differ.

TYPES OF ORGANIZATIONAL DIFFERENCES

- Management personality
- Customers
- Products
- Culture
- Location
- Profit
- Environment
- Quality levels
- Productivity
- Technology
- Core competencies

FIGURE 8.3
The ways organizations differ.

We realize that we covered these before, but we wanted to reinforce them in your mind.

The organization's mission documents the reason that an organization is formed. It should be broad enough to have a direct relationship to all activities that go on within the organization. A mission statement should change infrequently. It usually changes only when the organization decides to pursue a completely new market. It provides the overall boundaries that encompass objectives, goals, strategies, and visions.

The organization's vision statement is a view of a future desired condition or state of being. A vision should stretch the organization to be the best it can be.

The organization's values are things that an organization holds as sacred—things that should never be sacrificed. Much care should be taken when an organization selects its value statements to be sure they do not conflict with each other. Values are often interchanged with beliefs and principles. Management is held accountable for complying with the value and principle statements.

An objective communicates direction or intent that an individual or group has for the future. It is the desired end product of an activity or series of activities. It provides a basis for the goals. (*Example:* develop an organizational image for high service quality and reliability.)

A goal documents a desired quantified measurable result that the organization wants to accomplish in a set period of time to support its business objectives. (*Example:* increased sales at a minimum rate of 12% per year from 2010 to 2015 with an overall minimum average growth rate of 13%.) Goals should be specific rather than general so that there is no ambiguity.

PLANNING AND THE LEARNING ORGANIZATION

The single biggest change to the organizations culture in the twenty-first century is the transformation to a learning organization with its accompanying knowledge management system. The elusive search for the holy grail of learning and development has always been the creation of a learning organization. So how do organizations get there? Peter Senge popularized a holistic view of a learning organization in his book, *The Fifth Discipline: The Art and Practice of the Learning Organization* (1990).*

> [Learning organizations are] organizations where people continually expand their capacity to create the results they truly desire, where new and expansive patterns of thinking are nurtured, where collective aspiration is set free, and where people are continually learning to see the whole together.

---

* Peter Senge's *The Fifth Discipline* is his master work on the Holistic view of the organization using a systems perspective, and is divided into five parts. The first part is devoted to laying out the argument that we are the creators of our own reality, i.e., that the solutions to the problems that we face are within our reach, and that as life-long learners, we have the power to control our destinies. A major learning tool is the concept of "a Lever," or leverage points in a system, where the smallest efforts can make the biggest differences. (The so-called Butterfly Effect, or the sensitive dependence upon initial conditions). Senge also introduces the five disciplines of the learning organization, which are systems thinking, personal mastery, mental models, building shared vision and team learning. The entire work highlights systems thinking as the 5th discipline—the one which fuses the planning system into a coherent body of theory and practice. The book also contains a description of seven learning disabilities which are often responsible for organizational failure. Peter Senge, *The Fifth Discipline*, (New York: Random House/Doubleday, 1990/2006).

Countless pages have been written on this idea since that time and so there has been much greater awareness of what makes an organization a real learning organization. What do you have to do to reach this goal? The authors have seen all the successful learning strategies in practice. In this chapter we look at the main interrelated characteristics (identified by ASTD, the American Society for Training and Development) that are consistently present in the most successful learning organizations. These are (a) alignment to overall organizational goals, (b) efficiency, (c) effectiveness, and (d) sustainability.

- **Alignment:** It's important, right from the start, that your learning strategy is aligned to your organization's goals. You can set up highly efficient and potentially sustainable learning initiatives, but without real alignment they will never be properly effective.*
- **Efficiency:** Once you have aligned your learning initiatives, you need to get the most out of the limited budgets that you have at your disposal. Efficiency has been a major driver behind the adoption of new technologies and continues to be a major catalyst for change in both the private and public sectors. Ultimately, efficiency without effectiveness is pointless; both represent the key components in achieving a return on investment from your training investment.
- **Effectiveness:** So, to be effective you have to match agreed-upon learning needs with the right content and the right approach, at the right time. If you are not aligned, this is not possible. Combine effectiveness and efficiency and you have the key components of return on investment. Without effectiveness, it will simply be savings on investment (SOI). Effectiveness must lead to tangible benefits, which need to be broadcasted to all of the key stakeholders.
- **Sustainability:** The final key element is that your aligned, efficient, and effective learning strategy is set up in a sustainable way. The key is to build sustainability into the whole learning and development infrastructure. You do this by enshrining best practice in transparent

* Alignment to your organization's goals is the key to the success of your learning strategy. You can set up highly efficient and potentially sustainable learning initiative but without real alignment it will never be properly effective. ASTD research shows that: "BEST (Award Winning) Learning Organizations have formal processes to align short and long-term business strategies with competency, learning, and performance solution needs and priorities. The BEST map learning resources to competencies, individual development plans, jobs and corporate goals." ASTD 2005. In our opinion, it is against these criteria that all learning activities should be evaluated. *It's All About Alignment.* http://store.astd.org/Default.aspx?tabid=167&ProductId=1759 (accessed November 10, 2011).

processes and procedures, engaging the whole organization in taking responsibility for the continued success of the learning operation, and creating self-perpetuating networks of champions and supporters of learning within the organization.

Imagine you are driving an SUV on a camping vacation. You know that it is important to have a destination in mind, for this is your *long-range goal.* The destination alone, however, is not enough to get you there successfully, for you need to have a detailed map of instructions about which roads to take, when to make turns, the estimated distance and time, where you can stop for food and gas, along with gauges that tell you how much gas you have in your tank, and other warning systems to tell you if the engine gets overheated.

Now imagine that you are not driving the car alone, but instead you have 20 people doing different jobs simultaneously. The organization's CEO is at the steering wheel with a couple of board members looking over his or her shoulder, but four others are at each of the wheels making them spin; other people are looking out each window, reporting what they see to the driver, and someone else is in the back making sandwiches for lunch. It is going to take an impressive Annual Operating Plan to move this crew in the same direction.

This is the stuff of the Annual Operating Plan: which programs and management functions are going to do what, by when, and how much "gas" (money and person power) it will require. This level of detail is unnecessary in a Strategic Plan; in fact, it would clutter up the presentation of the long-range vision: the Strategic Plan focuses on the swimming hole at the camp you are going to, not which gas station to stop at along the way.

> Two sure ways to fail: Think and never do, or do and never think.
>
> **—Zig Ziglar, American author**

DEVELOPING A STRATEGIC IMPROVEMENT PLAN

The following are the five phases of the strategic improvement planning cycle:

- Phase I: Assessment
- Phase II: Vision Statements

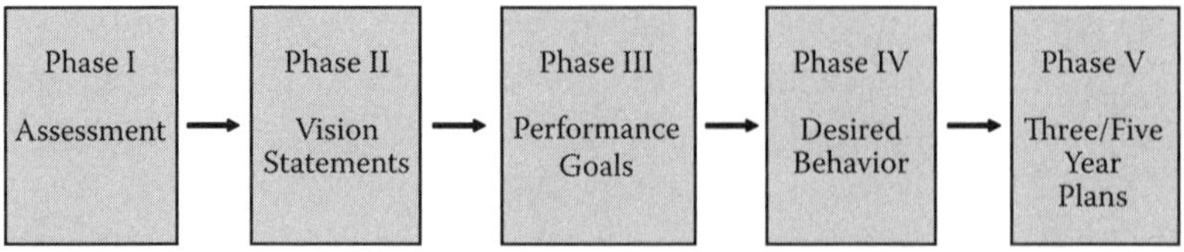

FIGURE 8.4
The five phases of the strategic improvement planning cycle.

- Phase III: Performance Goals
- Phase IV: Desired Behavior
- Phase V: Three- to Five-Year Plans

See Figure 8.4 for the five phases of the Improvement Planning Cycle.

Phase I: Assessment of the Organization

During the assessment phase, the key business drivers are defined, the present culture is evaluated, and an as-is view of each of the KBDs is developed. Key business drivers (KBDs), also called controllable factors, are the things within the organization that management can change that control or influence the organization's culture and the way the organization operates.

Phase II: Develop Vision Statements for Each of the KBDs

During this phase, the executive team develops a statement of how they would like the organization to be functioning related to each of the KBDs five years in the future. These first-draft vision statements are then presented to organization stakeholders for their comments and suggestions. Based upon these inputs, the set of 8–12 vision statements are revised and finalized.

Phase III: Develop a Set of Performance Goals

In this phase, the management team develops a set of performance goals by year for the next five years. The yearly goals are used to measure progress.

Phase IV: Defining Desired Behaviors

Improvement in culture and performance requires changes in the behavioral patterns of the management team and its employees. Because the team's behavioral patterns need to change before the desired results can be measured, a set of behavioral patterns that reflect the changes in behavior required by the 10–12 vision statements is identified.

Phase V: Preparing Three- to Five-Year Improvement Plans for Each of the Vision Statements

During this phase, a team will address each of the vision statements and define the problems related to the vision statement that the organization is facing today. Then they will make a list of the roadblocks that the organization will face in making the transformation. Once these are defined, the root causes for each of the problems and roadblocks will be identified. For each root cause, the team will select a tool or methodology to overcome the obstacle. A list of over 550 tools and methodologies can be found in Appendix B of this book. The team will also define the time required to implement the selected tools and methodologies.

ASSESSMENT OF TODAY'S PERSONALITIES

In developing our Strategic Improvement Plan, we need to define the "as is" status of the organization related to the following elements:

- Competitive position
- Core competencies
- Core capabilities
- Basic beliefs
- Customer satisfaction
- Employee satisfaction
- Quality management systems
- Successful and unsuccessful improvement activities
- The organization's commitment to improve

To do a complete assessment can be very time consuming and expensive, but when you take a look at the alternative, it is usually less expensive in the long run. If the organization is relatively advanced, most of this data should be readily available within the organization. If this is the case, it is best to take the time to do a comprehensive analysis and define any voids in the as-is organizational structure. The advanced organizations will typically already have benchmarked their competitors' products, defined and compared their core capabilities and competencies, calculated the customer satisfaction index, held focus groups with customers, conducted employee opinion surveys, and have an Organizational Master Plan that is actively being followed.

If this is not the case, a minimum analysis may be the right answer for your organization rather than completing a more thorough analysis in preparation for the three-year plan. This allows the organization to start the improvement process and modify the plan as additional data is made available. At a minimum, the assessment would include the following:

- Review of customer-related data
- Review of competitor-related data
- Review of the Strategic Business Plan and critical success factors
- Review of upper management measurements and performance
- Review of the organization's measurements and performance
- Review of quality systems audit data
- Review of past and planned-for improvement activities
- Review of grievances
- Private interviews with all top management to identify potential improvement opportunities or problems.
- Focus group meetings with middle management, first-level management, and employees to define the present personality of the organization, determine its compliance with business principles, and to identify roadblocks to change
- Perform a change resistance survey
- Conduct an improvement needs survey
- Conduct a key environmental factors survey

This type of assessment will provide a good definition of the as-is status. It will also identify many improvement opportunities that should be included in the Strategic Improvement Plan. In addition, the focus groups and the surveys will identify the differences in management and employee priorities. One of the surveys we used also provides a measure of the organization's dedication to improvement that can be used to evaluate the probability of successfully implementing the improvement process. In this case, an organization can be rated on a scale of 0–1,000. Zero indicates that the organization does not want to improve in any way; 1,000 indicates that the organization is extremely bad in all areas and that the maximum amount of improvement is desired in all areas. The average North American organization that we have worked with in implementing improvement processes is initially rated at 324.

The following is a breakdown of actions that should be taken based upon the improvement point count.

- **0–125:** A new improvement process will probably not be successful and should not be undertaken at this time.
- **126–240:** Before starting an improvement process, the group of people whose ratings fell between 126–240 should be exposed to other organizations that have made significant improvement, and a Malcolm Baldrige National Quality Award assessment should be conducted.
- **241–450:** Charge ahead. These organizations will make significant improvement in the next 3 years. Their return on investment should be greater than 8:1.
- **451–650:** A major effort will be required to meet the organization's goals in 3 years. Breakeven will not occur until the end of the second year.
- **Over 650:** It will take 5 years for the organization to meet its goals.

It is very important that the data collected during the private interviews, focus groups, and surveys be kept absolutely confidential. This is one of the best reasons for using a third party to do the assessment. The third party also provides an unprejudiced view of the organization, which is needed if the true as-is picture is to be defined. When individuals live in a problem situation for a long period of time, they become used to it and begin to consider it normal, rather than the exception that it really is. Do not rely only on the people who created the organization's personality and problems to define improvement opportunities.

Once the assessment is complete, the results should be reported back to management, and at a minimum, to every employee who took part in the assessment. If you ask anyone for their suggestions and/or to evaluate present status, you have an obligation to get back with them to show them how their data was used and reported to upper management.

HOW DO YOU CHANGE AN ORGANIZATION'S PERSONALITY?

Change is difficult under the very best circumstances. Western nations' change activities have been primarily driven out of fear—fear of losing market share, fear of losing jobs, fear of not making enough profit to keep stock prices up, and fear of failing. Eastern nations, on the other hand, are changing because they see an opportunity. It is too bad that most US organizations had to get into problems before they got serious about improvement. For example:

- Xerox and General Motors lost more than 30% of their market.
- IBM began posting record-breaking financial losses.
- The US government is so far in debt that there seems to be no way out. For the first time in history, the United States is classified as a debtor nation.

All organizations need to have a Strategic Improvement Plan, not just the losers. It's too bad that our leaders have been so slow in realizing this. Even Japan is looking at increased focus on improvement as corporations are downsizing. Lifetime employment in Japan is a thing of the past. (*Example:* Toshiba eliminated 5,000 jobs, and Fuji eliminated 6,000 jobs.)

What needs to be done to make improvement possible and long lasting? What needs to be done to bring about a change in the way we think, the way we talk, the way we walk, and the way we act? The following is an effective model for change.

1. Everyone must feel that change is necessary.
2. There has to be a common vision of how the change will affect the organization's environment.
3. Everyone should feel ownership in the Strategic Improvement Plan.
4. Management must change first and be the model for the rest of the organization.
5. Barriers must be broken down and removed by management.
6. The impact of the change must be openly communicated to each of the stakeholders.
7. Everyone needs to be trained to perform well in the new environment.

8. A measurement and feedback system needs to be put in place.
9. A risk-taking environment must be provided.
10. Directing and coaching must give way to leading.
11. Desired behaviors must be rewarded and undeserved behaviors changed or punished.

KEY BUSINESS DRIVERS/CONTROLLABLE FACTORS

Management has control over relatively few things. They do not control the economy, their customers, their competition, their suppliers, government regulations, the stock market, and so on. Management controls very little, but the organization is impacted by many things. The only things that management can change are the key business drivers that they control. If you want to bring about change in the organization, the key business drivers within the organization that impact the desired results must be changed.

An executive team should now be formed. This executive team should be made up of the highest officials in the organization (e.g., the president and all vice presidents) plus the key union leaders. The executive team should be limited to about 8 to 14 people.

The executive team now needs to define the KBDs that have the most impact upon the organization's performance. Most organizations have only 8–12 KBDs. Typical KBDs that impact organizational performance are as follows (KBDs with an asterisk (*) have predefined 12-level maturity grids already developed for them):

1. Measurement system*
2. Training*
3. Management and leadership methods*
4. External customer partnership interface*
5. Supplier partnership*
6. Business processes*
7. Production processes
8. Corporate interface to the divisions
9. Employee partnership
10. Research and development activities
11. After-sales service processes
12. Knowledge management*

PREPARING KEY BUSINESS DRIVER VISION STATEMENTS

It is very important to point out that there is a very big difference between KBD vision statements and the ones that are developed to define the vision of the organization as a whole. The difference amounts to this: the organization's short- and long-range vision statements define how the organization will look to the outside world, while the KBD vision statements define the internal workings of the organization. They clearly relate to the things that interest the employees.

The executive team should set aside two or three days to define the KBDs and develop a set of preliminary visions of how these processes will evolve over the next three to five years. To accomplish this, the meetings should be held off-site in a sterile environment where no one is interrupted by phone calls. The location should be selected far enough away from the work environment so that the executives will not return home in the evening. This session is designed as an executive team–building exercise, as well as a planning activity. The time during the day discussing visions is important, but equally important are the evening activities where the executive team gets to socialize and informally interact with each other. Typical evening activities can include the following:

- Decorating straw hats and judging the funniest, best looking, most creative, and so on.
- Having someone make a list of animals that each executive looks like and having the executives match the names with the list of animals.
- Small teams design paper airplanes to see which team can get their plane to fly the farthest.

The executive team should define the KBDs that impact the organization's performance by brainstorming to make a comprehensive list, and then consolidate it and prioritize it down to a maximum of twelve. Once the KBDs are defined, the executive team should define the present state of each of the processes. Twelve-level maturity grids have been developed for the seven most common KBDs to help the team quickly define the as-is state. Then they should discuss each process to define if it needs to be improved over the next five years. If the process needs to be improved (and most of them will), the executive team should define how it should change.

For example, one of the KBDs that always impacts performance is the management support/leadership process. During the assessment phase, selected personnel would have reviewed the following maturity grid to define which one best matches the as-is condition of the organization.

Key Change Area: Management Support/Leadership

1. Managers give orders. Employees are responsible for following them exactly without question. Management gets credit for all successes; employees are blamed for failures.
2. Managers give orders. Employees are responsible for following them exactly, but are allowed to question them. Employees are blamed when they do not follow orders.
3. Managers are responsible for results; workers respond to the directives of management.
4. Managers recognize the need for change. Recognition and rewards begin to be a part of the motivation process. Managers start looking for and praising people who do the right things, right.
5. Managers create a vision of the preferred future, which leads to group development of the *mission*. An organizationwide plan for achieving the mission has been developed. All managers are trained in systematic participative management. Teams are formed to work on problem-solving and improvement opportunities.
6. A continuous improvement process is launched; team building and problem-solving training are provided to everyone. Managers recognize the need to be process oriented. Progress has been made in building pride of accomplishment and self-esteem. Supervisors and managers are selected based primarily on their leadership ability.
7. Managers begin teaching, coaching, and working with their people on continuous improvement. Managers are treating quality and productivity as one. There are numerous examples of team building. An error-free performance standard is being used.
8. Management is working to change systems and processes, which their organizations have identified as barriers to achieving the organization's mission. "Management by walking around" is actively practiced. All employees are active members of a team. Employee opinion surveys are used by supervisors, and their teams, to improve.
9. Managers tailor their organizations to facilitate continuous improvement. Quality and productivity performance levels and improvement

projects are routinely reviewed with teams and individuals. All employees' output quality is measured and reported back to them.

10. Managers implement appropriate situational leadership concepts to stimulate groups and individuals in their groups in the implementation of a continuous improvement process. Managers are using statistical thinking. Teams are starting to set work standards. Promotions go to the people who prevent errors. An Organizational Master Plan that includes improvement activities is defined and understood by all.
11. Recognition and rewards clearly flow to those who are using a continuous improvement process. Firefighting is left to lower management and employees. Upper and middle management spend much of their time working with employees in their work areas, talking to customers, or doing long-range planning.
12. The culture includes the effective use of a continuous improvement process to continually improve quality, productivity, and employee morale. Workers are responsible for results; managers are responsive to their needs for achieving those results. Long-term quality goals are understood and supported by employees. Employees are setting their own time standards. The majority of management time is spent preventing errors.

This not only helps to define the as-is state, but also stimulates the executive team to think beyond the present boundaries and define the future desired state. After the executive team has brainstormed a list of phrases that best represents the future state of the specific KBDs they are discussing, they should prepare a vision statement for the specific KBD. This 5-year vision statement should clearly define the desired way the organization is operating 5 years from now. A vision statement should be written in the present tense. (*Example:* "We are effectively using participative management methods in all areas," not "We should be effectively using participative methods in all areas.") The vision statements should state what you are doing, not how you are doing it. (*Example:* "We are using statistical process control in all appropriate areas" rather than "We use P charts in the plating department.") The vision statement should cover the correction of today's problems, as well as future changes in the environmental condition. A typical example of a 5-year vision statement for management support and leadership is:

Management fosters an environment of open communication where opinions and suggestions are encouraged and valued: visions, plans, and priorities are shared throughout the organization.

Management provides the necessary time, tools and training for employees, which enables everyone to contribute their personal best toward the mission of the organization.

Teamwork is stressed: decision making is accomplished at the lowest appropriate level. Bidirectional feedback occurs on an ongoing basis to measure results and provide input for a continuous improvement process.

Organized Labor Involvement

If some of the employees in the organization are represented by organized labor, it is strongly recommended that the union leaders get involved as early as possible. We recommend that the appropriate union leaders became active partners in developing the KBD vision statements and the plan that will transform the organization. This will help to align organized labor's goal with the KBD vision statements. Early involvement of these key people in the planning process often slows down the process a little, but in the long run, much time will be saved as the plan will be much better and much more effectively implemented.

Stakeholder Involvement

The executive team develops the preliminary vision statements. These statements reflect the way management interprets the data they have and the way they picture the evolution of the organization's KBDs. But management is only a small part of the people who are affected by these vision statements. There are three more stakeholders who also need to influence these vision statements. They are: (1) the customer, (2) the employee, and (3) the suppliers.

Each of the executives should take the preliminary vision statements back to their organization and hold a series of focus group meetings with their direct reports, first-line managers, and employees. Each focus group should review all the vision statements to determine the following:

1. Is this the type of environment you want to live in?
2. Is this different from today's environment?
3. Do you understand the vision statement and what each word means?

4. How could it be improved?
5. Do you think it is achievable?
6. What would keep us from achieving it?
7. What can you do to help implement the visions?

Often these sessions are kicked off with the person who did the assessment reviewing the findings. This helps everyone obtain a better picture of the as-is condition.

Flip charts should be used to record all comments, negative or positive. This is an effective way to document the discussion and come to common agreement on key issues. Somehow things look different in print.

At the end of the focus group meeting, the executive will thank the group for their input and tell them that they will be reviewed with the executive team. The executive must point out that although all inputs will be considered, he or she cannot guarantee that they will be incorporated.

Procurement should ask the major suppliers to attend a focus group meeting where all the vision statements are reviewed, but most of the supplier focus group time would be directed at the supplier partnership vision statement. Marketing and/or sales should do the same thing for their major customers, with particular emphasis focused on present customer partner status and on the customer partnership vision statement. With both the suppliers and the customers, it is better to review the vision statement with too many, rather than too few.

This is one of the most exciting parts of the whole improvement process. In most organizations, it is the first time that management has ever asked the employees what type of environment they would like to spend their lives in. Even if the executive team only talks to 2 or 3 percent of the employees, the word spreads like wildfire throughout the organization. It is too bad that because most organizations have a major problem with employee trust, the typical comment is, "Looks great, but you have to prove it to me."

Preparing the Final Vision Statements

When the results of the focus groups are available, a second meeting of the executive team is held to develop the final vision statements. At this meeting, the executives present their teams' input and represent their teams' views of the desired future. Big arguments break out over small points;

that is, which synonym to use, which adjective to use, where the comma goes, and so on. After some agonizing debate, agreement is reached on a final group of vision statements. Each word in each statement has a common meaning to the members of the executive team. Unfortunately, these vision statements normally do not include all the suggestions made by the focus groups. In fact, different focus groups often have opposing viewpoints. For example, a group might object to the word *empowerment*, and another group loves it.

The outcome from this meeting is a new, final group of vision statements. In our experience, most of the final vision statements are very different from the preliminary vision statements. We have conducted many of these KBD visioning sessions and in every case the final set of vision statements has been different from the preliminary vision statement. This difference is very important. Even though an individual employee did not have his or her suggestion included, he or she is able to see that the executives changed their visions after they talked with the employees. It becomes very obvious that management is listening and really caring about the employees' opinions.

Organizations tap the full potential of their employees when meaningful, common visions are created jointly. Management's job is to promote these agreed-to, common visions. *Promote* does not mean to only talk about them or to support them. It means to *live* them, to *sell* them, to be enthusiastic about them. Confusion reigns supreme when management talks and writes one message, but acts and lives another. If management cannot live and act the visions, for heaven's sake, don't talk about or write them.

There is one bank in Arizona whose management felt strongly enough about their vision statements that they rented billboards on the route their employees traveled to and from work. On these billboards, the bank's mission and visions were posted. The message was not there for their customers. The customers would see the results of these visions. They were there for the employees. It showed the employees that management was firmly committed and that with the employees' help, they could bring about the major environmental changes that the vision pictured.

> Plans are only good intentions unless they immediately degenerate into hard work.
>
> **—Peter F. Drucker, American author**

SETTING PERFORMANCE IMPROVEMENT GOALS

The executive team should now define how it will measure success for the improvement effort. Every one of the executives has expectations of what should be accomplished in their function and the organization as a whole. To guide the improvement planning cycle, the executive team needs to focus on setting goals for only a few critical organizational measurements. For example, the following:

- Return on investment from the improvement process
- Customer satisfaction
- Response time
- Value added per employee
- Market share
- Error-free performance
- Cycle time
- Dollars saved
- Morale index
- Span of control

We suggest that the executive team select 3 to 6 organizational measurements around which the improvement process will be designed. The ultimate design of the improvement process will be greatly impacted by how aggressive these improvement performance goals are.

Next, the executive team will develop improvement goals for each of the measurements. To accomplish this, they will need to define how the measurement will be made. For example, if the executives decide that customer satisfaction is one of the critical measurements, it could be measured by a survey that uses a 1–10 scale (1 being poor and 10 being outstanding). The executive team could decide that they want a higher percentage of their customers to rate them between 8 and 10. The executive team then needs to define, by year, a goal for each improvement measurement. Using the same customer satisfaction example, if the executives knew that 60% of their customers rated the organization between 8 and 10 today, their goal for customer satisfaction might be as follows:

Year	Customer Rating between 8 and 10 (%)
0	60
1	65
2	70
3	80
4	85
5	95

If the organization does not know the customer satisfaction level at the present time, they could set a target of reducing the number of customers that rate them below 8. For example:

Year	Reduction in Ratings below 8 (%)
1	10
2	Another 10
3	Another 20
4	Another 40
5	Another 40
Minimum 5-year reduction	85

The same type of approach can be used for each of the other measurement goals. Figure 8.5 is a set of typical improvement performance goals that we developed for one of our clients. The performance goals defined in the Strategic Improvement Plan and the Annual Operating Plan must be in alignment with each other and in keeping with the vision statements.

	Year					
Measurement	**0**	**1**	**2**	**3**	**4**	**5**
1. Customer satisfaction	60%	65%	75%	80%	85%	95%
2. Return on investment from TIM	—	1:1	4:1	15:1	30:1	40:1
3. Average team return on investment	—	2:1	—	5:1	—	10:1
4. Defect rate improvement	—	2X	—	10X	—	100X
5. Value added/employee in $1000	45	50	65	70	75	80
6. New product cycle time in months	53	53	—	30	—	15

FIGURE 8.5
Typical high-level set of improvement goals.

DESIRED BEHAVIOR AND HABIT PATTERNS

The start of the personality change in the organization is developing a set of vision statements. If they are worthwhile and are embraced by management and employees alike, then the individual's feelings and thought patterns will begin to change. If the organization and the individuals involved are rewarded personally and socially as these new feelings are embraced, over time they will transform into normal behavior and habit patterns. For example, if part of your management support vision statement was to "empower your employees at all levels," this part of the vision could first be reflected in your employees as they begin to feel that they don't have to get management approval to take action on unplanned events. They would begin to gain confidence that they could make the right decisions in many cases without management's help. With time and continuing management support, they will begin to feel confident that they will not be hurt because they make a decision, and their behavior patterns will change. Now more and more, they will take the needed action, often telling management after the fact about the problem and how it was handled. They will start to come to management, explaining how they are going to correct the problem instead of asking management how to solve it. Positively reinforced behavior and actions become habits. At this point in time, these special patterns become a natural pattern. "It's just the way we do things around here; it's nothing special."

> Bring me your solutions, not your problems.
>
> **—H. James Harrington**

For every vision statement, a list of habits and behaviors that would exist in the organization if the vision was realized should be prepared. To accomplish this, the executive team may decide to focus on keywords or phrases in the vision statements, or the vision statements as a whole. Typical keywords or phrases that might be included in your vision statements are as follows:

- Empowered employees
- Customer-driven
- Adaptability
- Process focus
- Bureaucracy-free
- Streamlined operations

- Quality first
- Technology-driven
- Fast response times

Using *empowered employees* as a key phrase, the following is a list of some of the behavior and habit patterns that would be observed in an empowered work force:

- Self-managed work teams are used effectively.
- Wild ideas are encouraged and discussed.
- Unsolicited recommendations and suggestions are often turned in.
- Business information is readily available to all employees.
- Management defines the results expected, not how to get them.
- Failures are allowed and accepted as part of growth.
- Decisions are made more quickly and at lower levels.
- People manage their own resources.
- There is less second-guessing.
- People feel free to express their thoughts.
- People define their work process and time schedules.

Now the executive team needs to select key behavior and habit patterns and establish a way of measuring how they are changing within the organization. For example:

- Self-managed work teams could be measured by the percentage of people that are part of these teams.
- The degree that wild ideas are encouraged and discussed could be measured by reviewing brainstorming lists to determine what percentage of the items stretch the imagination.
- How often unsolicited recommendations and suggestions are turned in can be measured by the number of performance improvement ideas and suggestions that are turned in per eligible employee.

Using this type of thought pattern generates a very extensive list of behavior and habit pattern measurements, many of which are not being measured in most organizations. The executive team should review this proposed list of measurements and include many of them when creating their measurement plan.

IMPROVEMENT PLANS FOR EACH KBD

Key Business Driver Improvement Plans are plans designed to transform the organization from the as-is state so that it is in line with the relevant KBD vision statement. They are usually 3 to 5-year plans.

Organizations that want to eliminate the piecemeal, flavor-of-the-month approach to improvement are stepping back and looking at all their improvement options before committing to a course of action. It takes time up front, but it saves total cycle time, cost, and effort over the life of the project. In addition, it produces much better results. Properly designed, it will create an organization that is creatively bringing out the best each employee has to offer. Work becomes a rewarding, enjoyable, exciting, experience. The environment promotes a team spirit without taking away the individual's sense of accomplishment, achievement, and self-esteem. The excitement of belonging and achievement creates an electrifying air that snaps like lightning bolts among management and employees alike, breathing new life into the entire organization. It eliminates the flavor-of-the-month approach and the employees see for the first time that management looks at improvement as an important ongoing part of the business.

David T. Kearns, past chief executive officer of Xerox, stated, "Our primary motivation in applying for the National Quality Award was to find out how we could improve. Sure, we wanted to win. But we wanted to learn even more. We spent the last year using the award process to identify areas in which we can improve. And we're using what we learned to kick off a second 5-year plan for quality improvement at Xerox worldwide."*

Yes, a multiyear improvement plan (Strategic Improvement Plan) lies at the heart of every successful improvement activity, and not just for the large corporation. We reviewed Globe Metallurgical's QEC Continuous Improvement Plan. Globe was the first small business to win the coveted Malcolm Baldrige National Quality Award. Globe's continuous improvement plan contained 96 different objectives, with multiprojects for most of the objectives, with target completion dates distributed over a two-year period.

---

* David T. Kearns interview by Frank Voehl, Malcolm Baldrige Award Headquarters, Gaithersburg, MD, April 2, 1997.

Factors Impacting the KBD Improvement Plans

You may question why we are talking about a 3 to 5-year KBD improvement plan instead of just a 5-year plan. We have found that many of our clients, when they are just starting to prepare a long-range improvement plan, have a hard time defining the activities that they will be doing in years 4 and 5. In these cases we recommend that they do only a 3-year plan. At the end of the first year, fill in the plan for the fourth year. At the end of the second year, fill in the plan for the fifth year.

An organization must consider many factors before finalizing each of the KBD Improvement Plans. It can be divided into two categories: (1) impacting factors and (2) influencing data.

- Impacting Factors
  - Organization's mission
  - Organization's values
  - Environmental visions
  - Performance improvement goals
  - Future Business Plans
  - Desired behavior and habit patterns
  - Improvement/quality policies
    - Other projects
    - Resources
- Influencing Data
  - Opinion surveys
  - Independent reviews
  - Customer feedback data
  - Competitive performance
  - Poor-quality cost
  - Today's problems
  - Environment
  - Improvement tools

Certainly the environmental influencing data can have a major impact on the final KBD Improvement Plan. Included in these considerations are these items:

- Technologies
- Standards
  - State and federal government standards
  - World-body standards

    - Self-imposed standards
    - Government regulations
- Desired pace of change
    - Changing customer expectations and needs
    - New product releases to expand features and service
- Competitive environment
    - IT's impact on the processes

The proliferation of improvement tools has certainly increased the complexity of the improvement planning cycle. For example, consider Crosby's 14 Steps, Deming's 14 Points, Feigenbaum's 10 Benchmarks of Quality Success, and hundreds and hundreds of others, all similar, but still all different. What was the result? There are now more than 1,100 different improvement tools available for the organization to select from. (See Appendix B for the list of some of the over 1,100 different improvement methodologies.)

With so many methodologies and tools, each organization has a wide selection to choose from. This can be a blessing and a curse. It is a blessing because you can get just the right one for your situation; it is a curse because with so much to choose from, it is hard to make the best choice and you can choose too much. It is a lot like a boy going into a candy store and there is no one behind the counter. He just helps himself to all the candy he wants and goes out behind the barn and eats it all. In just a little while he gets very sick, and based upon that experience, he concludes that candy is not good for him.

We have seen presidents of organizations do the very same thing; they get disappointed in an improvement tool because they want it done too

FIGURE 8.6
KBD Improvement Plan influencing and impacting considerations.

much and too fast. As a result, they believe that their chosen methodology is not good for the organization. As a result, ...the organization because they start too many things at the same time without the resources to get them done, in addition to the normal activity, which drives the organization into future shock. This causes the organization to fail to realize the benefits that it should receive from the improvement activities.

> Don't try to do too much improvement or you will not get any improvement.
>
> —**H. James Harrington**

Which ones are right for you? Figure 8.6 shows the considerations that influence and impact the KBD Improvement Plan.

Difference between Planning and Problem Solving

We find there is a lot of confusion between planning and problem solving. North Americans and Europeans are good problem solvers, but they hate to plan. Planning sessions continuously flow over into problem-solving sessions. It is important to separate planning from problem solving if the KBD Improvement Plan is to be completed in an expeditious manner. Planning is upper management's responsibility. Problem solving is the responsibility of middle management, first-level management, and the employees. Look at the difference between the two:

Planning	Problem Solving
Define directions	Define solutions
Identify change areas	Implement changes
Assign resources	Use resources
Identify needed action	Take action
Highlight symptoms	Find root causes
Look at big picture	Focus on single issues
Short-term cycle	Long-term cycle

Developing Individual KBD

The individual *Key Business Driver Improvement Plan* should transform the organization's culture to the desired level. The combination of all the KBD Improvement Plans make up the Strategic Improvement Plan. They are usually 3- to 5-year plans.

The executive team should look at each vision statement and develop a plan to transform the organization's environment over the next 3 to 5 years in keeping with the vision statement. Often the executive team will assign one or two executives to head up a team of managers and employees to develop the improvement plan for each of the KBDs. This KBD Improvement Plan must provide a logical transition from the as-is state to the desired future state as defined by the vision.

Transition is defined as an orderly passage from one state, condition, or action to another. An effectively planned-for transition has the following characteristics:

- It is not abrupt.
- It does not create morale problems.
- It does not have schedule slippage.
- It is not uncontrolled.
- It is not an unplanned occurrence.
- It does not create customer complaints.
- It accomplishes desired results without rework.

The KBD Improvement Plan team will then generate a list of today's problems related to the KBD that is being planned for and a list of roadblocks that will impede the change of state. When this is complete, the tools that impact the KBD under study will be reviewed. For example, in the case of management support and leadership, about 44 of the 1,100-plus improvement tools are the most used to bring about this transformation. Typical tools are the following:

- Area Activity Analysis
- Self-Managed Work Teams
- Communication Skills
- Effective Listening
- Nonverbal Communication
- Inspiring, Leading, and Modeling
- Management by Walking Around
- Performance Planning and Appraisal
- Suggestion System
- Team Building
- Natural Work Teams

The appropriate improvement tools and the list of problems and roadblocks are then analyzed to determine which tools will be used to correct which problems. In many cases, different tools are effective on the same problems. The KBD Improvement Plan team must study these interrelationships to determine which tool provides the best combined results in their particular environment. Once the KBD Improvement Plan team has selected the appropriate tools, a tactic-level implementation for each tool will be prepared and an individual assigned the responsibility for ensuring the plan is implemented. At this point in the planning process, priorities are not given to individual tools unless there is some type of interdependency. Figure 8.7 illustrates a typical Gantt chart for an individual KBD vision statement.

You will note that the timing is month 1, 2, 3, and so on, not a month of the year (i.e., January, February, December). This is done because we are not scheduling the activity to be done by a specific month. The purpose of the chart is to show how long the project will take to complete and the interdependence among activities.

The KBD Improvement Plan team should then review the performance improvement goals to identify which measurements are impacted by the specific KBD change plan. The KBD Improvement Plan team should then evaluate how much improvement the specific KBD change plan will have related to the affected performance improvement goals. Although a

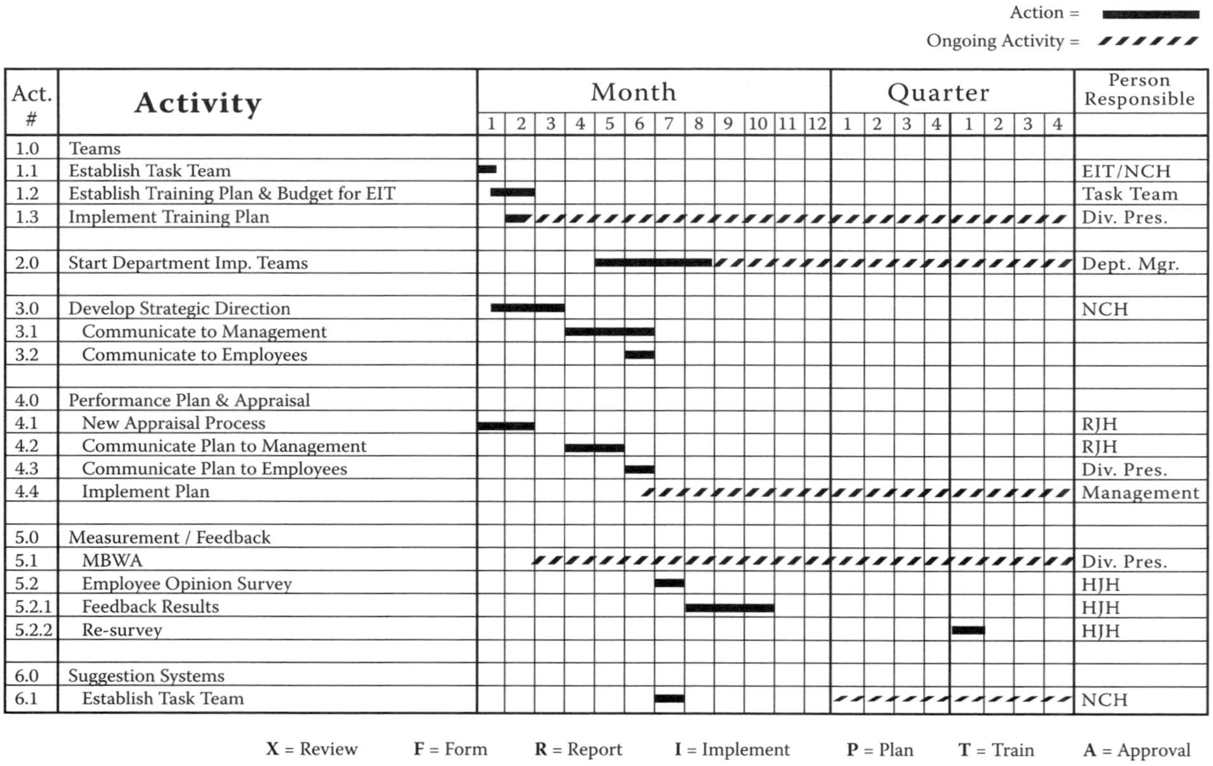

FIGURE 8.7
Individual KBD Improvement Plan.

number of KBD change plans can impact one measurement, the total sum of the impacts should add up to at least the minimum goal for each measurement. If this is not the case, the individual KBD Improvement Plans need to be improved.

The KBD Improvement Plan team should also evaluate each of the specific environmental change plans to ensure that they will be conducive to the desired behavior and habit patterns previously defined. If the individual KBD Improvement Plans do not meet this test, they will need to be modified.

THE STRATEGIC IMPROVEMENT PLAN CHECKLIST

All of the following items should be considered in developing an improvement plan for the organization. Now is the time that the executive team should see if the improvement plan includes the following items:

1. A long-term vision.
2. A vehicle for all managers and employees to understand and buy in to the plan and vision and to keep them informed.
3. Focus on continuous improvement and prevention.
4. Focus on the customer and customer-related measurement systems.
5. A way to provide each individual with a chance to excel and be recognized for individual excellence.
6. Management setting the example for the employees.
7. Focus on the processes that allow errors to occur.
8. A management system to guide the improvement process.
9. A means whereby all employees and managers will become actively involved.
10. A way to overcome all potential barriers.
11. Provision for the three necessary ingredients for error-free work: training, tools, and time.
12. Use of teams.
13. Effective job training and employee certification.
14. Process characterization and process capability studies.
15. A process flow and layout that will minimize in-process and final goods inventory.

16. A supplier partnership process that will allow materials and parts to go directly to the line.
17. Output quality feedback to all employees.
18. Provision to keep all managers and employees informed about progress.
19. Identification of those responsible for measurement and how progress against the plan will be measured.
20. Identification of outside dependencies, if any (i.e., manpower, funds, higher priorities).
21. Education—included for each item where appropriate.
22. Consideration of the as-is assessment recommendations.
23. A reasonable, overall schedule that can be accomplished within given resource constraints.
24. Assigned responsibilities and a timeline for each event.
25. Provision for the rewards to reinforce desired behavior.
26. Provisions to ensure that it meets the magnitude and schedule defined in the performance improvement goals.
27. Provisions to ensure that it will create the desired behavior and habit patterns.

MAKING THE IMPROVEMENT PROCESS WORK

The problem that we have had in the West is not what we do—it's how we implement it. Almost all the improvement tools that individual organizations have tried are good under the right conditions. Unfortunately, in most cases, we have done the right thing, but did it very ineffectively, thereby minimizing the potential gains from the activity. Literally billions of dollars have been wasted in the West training employees to do things that they are not using. The key to successful implementation of the improvement process is an excellent change management process.

> Create a definite plan for carrying out your desire and begin at once, whether you are ready or not, to put this plan into action.
>
> **—Napoleon Hill, American author**

SUMMARY OF STRATEGIC IMPROVEMENT PLAN

In this chapter we defined the controllable factors, also called KBDs, that the organization can vary to change the culture and the way the organization performs. For each of these KBDs, we developed a set of vision statements that define how the people within the organization would view the KBD if the desired changes were implemented. For each of the vision statements, we defined the problems and improvement opportunities we have at the present time and the road blocks that the organization would face in making the transition. Using the 1,100 plus tool kits we analyzed each of the problems, improvement opportunities, and roadblocks to define the tools that were best to make the required changes. Because the focus of this activity is to modify the way things are done within the organization, the tools and methodologies will usually require that a change in the behavior of the management and the employee will be needed.

We also defined ways to determine if the behavior modification techniques were working. We then developed a list of tactics required to use each tool and the time it would take to do the tactic.

The strategic improvement planning approach applies to both big and small organizations. We have used it with organizations as small as sixty-five people and with those who had thousands of people. Typical organizations that we have helped use this approach include

- US Air Force Space Systems Division
- Earle M. Jorgensen Company
- Martin Marietta Manned Space Systems
- Hewlett Packard
- LTV Missiles and Electronic Group
- Lambda Electronics Inc
- Eastern Space Missiles Center
- Allied Supply Company
- Arizona Department of Transportation

When you stop improving, you start to step backward because your competition is improving.

—H. James Harrington

9

Developing the Strategic Plan

If you don't know where you are going, you won't know when you get there.

—H. James Harrington

COMBINING THE STRATEGIC BUSINESS PLAN AND THE STRATEGIC IMPROVEMENT PLANS

We have now developed a set of Strategic Improvement Plans and Strategic Business Plans both to the tactics level. But the Strategic Plan is not complete until the two Strategic Plans are combined and prioritized.

There are a number of things that should be considered when the individual plans are combined, not the least of which are the performance improvement goals that were developed earlier. The executive team should review the individual plans to define which activities impact each of the performance improvement measurements and schedule the activities so that the performance goals will be met. Often you will find that there are multiple strategies or tools selected to address a single problem. Look at these very carefully to understand the impact that each of the strategies or tools will have on the item. Often you will find that you can get the desired results without using all of the strategies or tools that were directed at the item. We have seen times when there were two different strategies and three different tools directed at the same problem, and by just selecting the *right* tool, the item was brought under control. When you have multiple strategies or tools directed at the same item, install them far enough apart so that the impact of each is measured before the next one is implemented because it may not be needed. See Appendix C for an example of how to select just the tools you need.

Other things that should be considered in scheduling the activities in the Strategic Plan are

- Available resources
- Other activities going on within the affected area
- Holidays and vacations periods
- Seasonal and new product workload fluctuations
- Interdependencies
- Organized labor involvement
- Change management timing
- Impact of organizational change management

If, when all the constraints are considered, the Strategic Improvement Plan will not bring about the desired improvements as defined in the performance improvement goals, then either the goals or the plan should be modified.

The biggest single mistake most organizations make is in trying to implement the improvement process too fast. Great care should be taken to balance the improvement effort and resources required with the other activities going on within the organization. Most tend to overcommit to the improvement effort at this point in time. During the planning cycle, management gets very excited about all the improvement opportunities that are identified and their impact on the overall performance of the organization. They want to do them all right now. Why wait a year before the problem is assigned to someone to fix? In fact, it is much better to be conservative during the first year rather than too aggressive. It is our experience that organizations already feel that they have a workload that is 110% of their work force capability.

When the improvement process starts, there is an increase in workload. In some areas, the increased workload can go up as much as 30%. The total workload typically does not drop below its original value for 6–12 months. To offset this short-term, peak workload, consultants and temporary employees should be used. As employees are freed up from their present jobs because of the improvements made in the process, they should be deployed to replace the consultants and temporary personnel.

It is important to realize that along with change often comes a short-term negative impact on the organization. For example, if an organization was

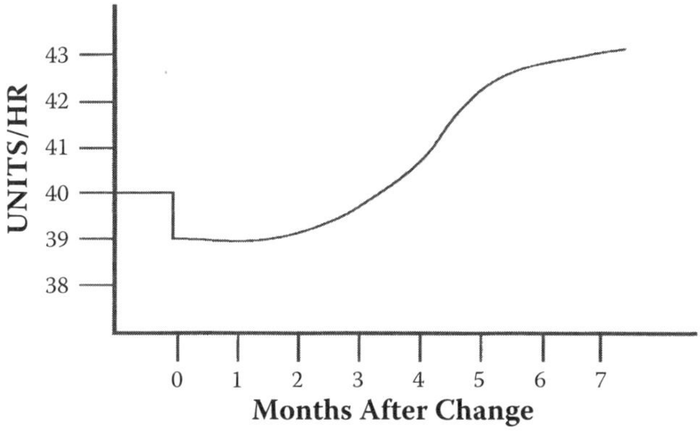

FIGURE 9.1
Impact of typical change on output.

producing 40 units per week and they started holding one-hour weekly team meetings, the output would automatically drop to 39 units per week (see Figure 9.1). Psychologically, to do otherwise would be to admit that you were not giving a fair day's work for a fair day's pay. In other cases, it occurs because the employees need to be trained and learn new methodologies. In these cases, productivity can drop as much as 30% during the training and familiarization phase. Typically with a team, the negative impact gradually disappears over a 3-month period and they are able to produce 40 units per week and still find time for the one-hour team meeting each week. As the team meetings continue past the 3-month point, both the output and quality of output usually surpass the original starting point.

> The one who figures on victory at headquarters before even doing battle is the one who has the most strategic factors on his side.
>
> **—Sun Tzu, Chinese general and author**
> *The Art of War*

CREATING THE STRATEGIC PLAN

Now we have briefly covered how to generate the two halves of the Strategic Plan—the Strategic Business Plan and the Strategic Improvement Plan. But do we have two halves that make a whole? Are there voids or overlaps in the two plans? If we work with them as two separate entities, there will be a lot of competition for people and other resources. Which plan or part of the plan has priority? Are there any interdependencies between the two plans? Which plan gets credit for the organization's success and which one gets the blame if things don't work out well? In reality, you should not have two separate plans; they need to be combined together so the total organization is singing the same song.

As a way to help you prioritize the activities, we suggest that you make a matrix of seven columns. In the left-hand column, list all of the strategies, tools, and methodologies defined in the two plans (see Figure 9.2). The headings of the other columns should be as follows:

- Column 2: Employee Impact
- Column 3: Dollars Saved
- Column 4: Business Impact
- Column 5: Customer Impact
- Column 6: Total
- Column 7: Suggested Year to Implement

Rate every block in the chart from 1 to 5 as follows:

- 1 = Very low impact
- 2 = low impact
- 3 = average impact
- 4 = high impact
- 5 = very high impact

Strategies, Tools, and Methodologies	Employee Impact	Dollars Saved	Business Impact	Customer Impact	Total	Improvement Year
Lean	3	2	3	1	9	1
Knowledge Management System	4	2	4	2	12	3
Operating Manual	3	2	2	1	8	2
Team Training	4	2	3	1	10	1
Self-Management Teams	5	2	1	0	8	4
Organizational Alignment	4	3	5	2	14	1
Supply Chain Management	1	3	4	2	10	2

FIGURE 9.2
Prioritizing Strategic Plans.

In this case, the higher the total, the bigger the impact on the organization. In most cases, the high-impact items get scheduled early in the cycle, but not in all cases. In some cases an item cannot be implemented until other conditions are completed first.

Using the matrix as a guide, a team of knowledgeable people can develop a Five-Year Strategic Plan that defines, by quarter, when each strategy, tool, or methodology should be started and completed. Both dates are important.

Figure 9.3 is a typical example of a Strategic Plan that combined the Strategic Business Plan and the Strategic Improvement Plan into a 3-year Gantt chart. You can see at this stage that we have taken into consideration things like summer vacation, holidays, strategic planning cycles, and so on. A major portion of this plan is the supporting change management plan that will help ensure the smooth implementation of the individual process changes. The whole concept of change management needs to be well understood by the executive team and the people who implement the plans, and integrated into the Organizational Master Plan.

SUMMARY

The piecemeal approach to improvement usually produces results, but not the best results. To become a winner, or to stay a winner, organizations

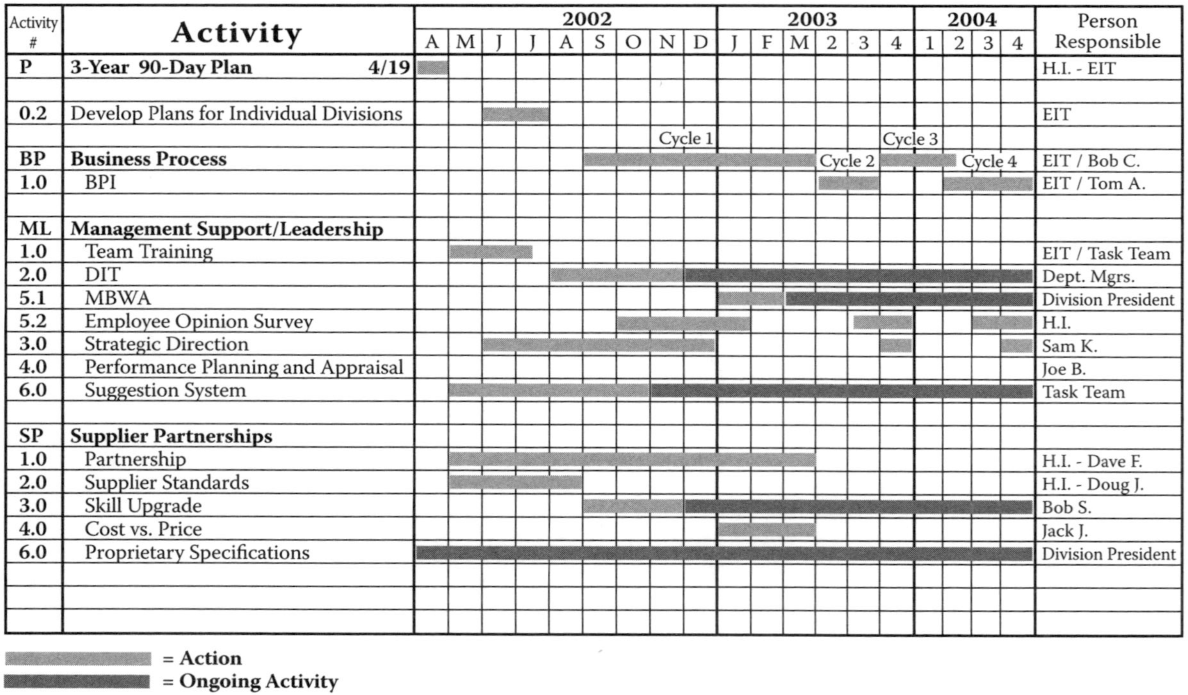

FIGURE 9.3
Combined 3-year Strategic Plan.

need to define how they want the organization's environment to evolve over the next five years, by preparing a series of vision statements that define the future desired state. Once their direction is defined, they can design the improvement process that will uniquely meet their transformation needs.

When the Strategic Business Plan and the Strategic Improvement Plan are completed, they must be combined, thereby creating the organization's Strategic Plan.

Organizations can no longer react to the latest improvement fad. They must consider all of the options that are available to them and then patiently implement them over a period of time so there will not be a negative impact on the organization's performance. Frequently, in their enthusiasm to implement the change process, organizations overcommit themselves, taking away resources that are needed to maintain their output.

Now let's look at how different types of organizations implement their improvement process.

Tool Selection

Losers always look for the latest fad.

Survivors select one guru approach and hold to it.

Winners look at all options and their impact on the organization and combine approaches to make a new one that is designed just for them.

Planning for Improvement

Losers plan to implement the tool of the day.

Survivors use a standard, proven approach based on what other organizations have found successful.

Winners use a highly customized plan that will be implemented over 3 to 5 years.

Payback on Improvement

Losers do not measure return on investments. Measurements are focused on activities, not on results.

Survivors look for long-term payback, but their primary emphasis is still on activities.

Winners look for the improvement process to pay for itself as it goes along, and measurement systems are established early in the process to measure return on investment.

Use of Visions

Losers: Top management develops an overall future general vision statement that describes how the organization will be perceived in the future.

Survivors: Top management develops an overall general vision of how the organization is perceived. Management then widely communicates and discusses the vision statement with the employees to ensure that they understand the vision statement and understand that they must be part of the change process.

Winners: A series of vision statements that relate to the organization's internal key business drivers are developed jointly by management and employees.

Purpose of the Improvement Plan

Losers: The purpose of the plan is to reduce customer complaints and the results are used as a marketing tool.

Survivors: The purpose of the plan is to stay competitive.

Winners: The purpose of the plan is to improve the organization's value as viewed by all of its stakeholders.

You can only be the best when you have tapped the full potential of your people and they are working for a common vision.

—Anonymous

A Strategic Plan has two parts—one that is directed at growing the business and another one that is directed at improving the way you do business.

—H. James Harrington

10

The Operating Plan

The Operating Plan turns yesterday's dreams into today's realities.

—H. James Harrington

OVERVIEW

The Operating Plan makes the jump from philosophy to reality. It is a commitment to get the job done very quickly. It's quantifiable and it's concrete. And there is no question how to determine whether you've reached it. It takes the goals that make up the Strategic Plan and breaks them down into subgoals that are short-term oriented. Subgoals are divided into activities that are assigned to individuals. Things begin to happen fast once you know exactly where you want to go and have a "roadmap" on how to get there.

Too many people equate the Operating Plan with budgeting alone. We believe that the budget is just one part of the Operating Plan and it is the result of many more important things that need to be addressed before we start to put dollars on spreadsheets. Before you start doing any nitty-gritty budgeting for your annual plan, here are the crucial first steps:

1. Review your organization's Strategic Plan to find out what activities need to be addressed during this cycle.
2. Define what goals need to be met to be aligned with the Strategic Plan. Establish just a few major goals for the next year.
3. Have each function define its goals and what its dependencies are.

4. Define the assumptions on which the Operating Plan will need to be based.
5. Benchmark your industry to be sure your goals are in line with your competition's performance.
6. Be sure that your goals are ahead of your changing internal and external customer requirements.

> Planning is bringing the future into the present so that you can do something about it now.
>
> **—Alan Lakein, American writer**

Although implementing the Strategic Plan, which includes the Strategic Business Plan and the Strategic Improvement Plan, is the key step in the Strategic Planning process, it may prove to be difficult to implement it in the Operating Plan. Much creative energy and time has been expended in processes, such as developing a mission, vision, and goals, and actual adoption of a plan; sometimes the interest of those involved in the process may begin to diminish. However, a great plan with good creative ideas and goals without a specific annual planning implementation process is useless. Implementation of the Strategic Plan is done incrementally. In other words, Strategic Plan implementation is done through the Operating Plan which describes the short-term strategies and resources that define how the overall Strategic Plan will be implemented during any given operational period (often the organization's fiscal year). The Operating Plan serves as the basis and justification for an operating budget request. It should describe the following:

- Where the organization is currently in implementing its Strategic Plan
- Which portions of the Strategic Plan will be implemented during the operating period over the next 12 to 18 months
- Where the organization will be at the end of the Operating Plan period
- How the Operating Plan will be implemented (which units and people are responsible?)
- How progress will be measured and reported (what are the department performance indicators and what are their improvement targets?)
- How key stakeholders will be informed and involved

PROLOGUE

Unfortunately, there is a myth that identifies Operating Plans purely with start-ups. That's particularly common in the United States. Because of that myth, along with inertia, putting out fires, and other related reasons, a lot of organizations miss out on the opportunity to manage themselves in a better manner. As an owner or manager of a small, medium, or large business, can you afford not to plan? Do you leave the Operating Plan to the large business organizations, and let your business depend on reacting to events? Or do you want to plan for priorities, and manage your growth proactively as part of a detailed Organizational Master Plan for success. That's a leading question, and of course the answer is obvious, especially for those organizations that we would call winners.

Many call it an Annual Plan, Business Plan, an Annual Operational Plan, or just the Operating Plan; the name doesn't matter as much as the management of it. While these kinds of plans are common in larger enterprises, they are surprisingly rare in small and medium-sized business because small organizations do not have the discipline of execution that many of the larger organizations do.

Charles Knight, chairman emeritus of Emerson Electric put it this way:*

> At Emerson,† management is about performance without compromise—achieving superior results consistently over the long term. For decades, we've organized and operated the company to achieve winning results. Between the mid-1950s through 2005 [a fifty year span] Emerson reported annual increases in earnings, earnings per share, and dividends per share. This record is matched by only a handful of companies in the world, and is unmatched by any that make products or provide services comparable to ours.

---

* Charles Knight, Performance without Compromise: How Emerson Consistently Achieves Winning Results (Boston, MA: Harvard Business Press, 2005), 1–4.

† Emerson is a $25 billion global manufacturing company offering a vast array of products and services that bring technology and engineering together to create innovative solutions for the benefit of their customers. They grew from a dozen divisions generating revenues under $1 billion to over 60 divisions today. They used the Management Planning system to transform the organization from a technology follower to a technology leader, and to reinvent their manufacturing operations for global competitiveness. They have been able to use their Master Planning system to restructure the company to respond to emerging trends in customer behavior. Over the years, they have completed more than 200 acquisitions with an investment of over $10 billion. Most impressively, they have been able to do all this and maintain an executive turnover rate of less that 4%.

How do we do it? Actually, there's no mystery. We have a dynamic approach to running the company that emerged in the 1950s and continues to evolve. We call it the Management Planning Process and (the resulting Master plan) is the key to performance without compromise. The annual management (planning) process helps us develop superior strategies at all levels of the company, and keeps us on track as we execute them each year with discipline and intensity. In this way, we avoid one of the most common troubles of corporations, which often struggle because they don't execute their strategies. For us, the (annual planning) process provides the link between strategy and execution. It establishes a common set of values. It provides a structure for setting priorities, evaluating results, and solving problems. And it establishes shared ways of thinking and communicating for ease and speed of execution.

The benefits of the Operating Plan are as follows:

1. **Guide your growth:** Your business will grow, or not, depending on a lot of different factors, including overall economic trends, location, specific market needs, hard work, and other elements. Businesses that plan do it to guide and influence their growth, so that they move proactively toward defined objectives rather than just reacting to business events.
2. **Manage priorities:** Strategy is focus. Allocate resources where they will do the most good. Work toward your strengths and away from your weaknesses. Develop the organization by doing the most important things, according to your long-term objectives.
3. **Assign responsibilities:** An Operating Plan gives you a place to develop organizational responsibilities.
4. **Track progress:** Think of the Operating Plan as a *global positioning device* for business. With an Operating Plan, you can track your progress toward goals, measure results, and manage the business. Without an Operating Plan, how do you tell whether or not you are moving in the right direction each quarter? And against what do you measure?
5. **Plan for cash:** Profits are not cash, and cash is not intuitive. You spend cash, you don't spend profits. However, businesses don't plan well for cash, and they need to do so. That may not sound strategic, but it is. It is also the core of the Operating Plan. Whatever else, you have to plan for cash.

The Operating Plan includes some or all of the following main points:

- *Focus on main priorities:* Planning assigns priorities. Of the whole range of possible market segments, and the whole range of services and possible sales and marketing activities, which are your main priorities?
- *Specific responsibilities, activities, deadlines, and budgets:* They are the bricks and mortar of the Business Plan, and are critical to business success.
- *Financial plan:* One of the most important outputs from a Business Plan is the financial plan, which of course hinges on cash flow. A business needs to stress its priorities by making sure it gets the right amount of money.

> The successful leader must plan his work and work his plan.
>
> **—Napoleon Hill, American author**

A team of managers usually works together to lead the whole organization to a key goal such as a breakthrough goal, strategy execution, or the resolution of a systemic constraint of the business system. Developing the Business Plan is considered to be one of the most difficult components of the organization's master planning process. Many sources quote failure rates of over 50% for most organization's master planning cycle projects. *The Economist** offers even more pessimistic estimates, suggesting that as many as two thirds of all projects come up short of expectations.

The Harrington Institute has conducted studies on projects over a period of several years and worked with senior managers in the client organization to develop their plans. Key assumptions were verified. Strategic Implementation Plans were developed for key functions and projects. Cross-functional issues were resolved and organizational consensus developed. Review mechanisms were put in place to monitor progress and ensure essential inputs. A robust operating planning process was developed and actual business performance has been largely in line with the plan. Our client organizations were able to achieve their best performance

* See the *Economist* article "Overdue and over budget, over and over again" (June 11, 2005).

in the years following the planning exercises and deployment, using the Organizational Master Planning method* described in this book.

On the other hand, our experience and research revealed three key reasons for failure of management planning projects:

- *Wrong commitments:* The plan's commitments are delivered, but we missed the goal.
- *Weak commitments:* The commitments are not well defined.
- *Broken commitments:* The commitments are not delivered.

> Unless commitment is made, there are only promises and hopes; but no plans.
>
> **—Peter F. Drucker, American writer**

With *right commitments*, *strong commitments*, and *all commitments delivered*, the organization's master planning cycle will be successful. The right commitments deliver those vital few results required to bring the management project to the goal. Unfortunately, managers often build their operational plans using the brainstorming and voting methodology. In these cases, the plans are often expressions of the organization's politics instead of reflecting the use of high-quality management planning tools. Unfortunately when dealing with system change, this approach has little to no chance to deliver the action plans required for creating long-lasting system change.

Instead, systematic methodologies must be rigorously applied to arrive at the right commitments. Successful operating planning is based on a deep understanding of the business system that management wants to lead to the goal of the Organizational Master Plan. Thus, deep research needs to be conducted for all obstacles to be overcome on the way to the goal. Finally, rigorous logical testing of the Operating Plan is needed before it is released for implementation and execution. Well-defined commitments consist of the following five elements, each of which must be defined correctly:

* The Organizational Master Plan is an invaluable strategic business tool for any type of organization, whether it's a biotechnology company, bank, university, or church. It's important that your facility Organizational Master Plan reinforce your overall strategic plan. The Master Plan involves the long-term design and planning of an organization and its facilities, be it a building, group of buildings, or an entire site. Typically, it involves the design of a first phase in the context of one or more future phases, all the way to a full build-out. Depending on how an organization grows and transforms over the years, the Master Plan may involve several options to provide flexibility to accommodate change and growth, meet current and future space needs, and support the foundation of the Strategic Plan.

- The owner: The person who commits to deliver
- The deliverable: What will be delivered
- The purpose: Why that commitment is vital for the goal
- The due date: The date when delivery is required
- The acceptable effort: Consists of the resources and budget available and needed

Making sure that all commitments in an Operating Plan are the right ones, that they are strong, and that they are delivered is top management's responsibility. Therefore, top managers of successful management projects do not fall into the "delegation trap." They own every element in the Operating Plan and don't delegate responsibility for it because they know that success lies in the execution. Charles Knight of Emerson Electric puts it this way:

> As potent a management tool as planning is, it has no value if it cannot or will not be executed. At Emerson, we take execution just as seriously as we take planning our annual cycle of follow-up and control helps ensure that we put our plans into effect and track progress against them. We invest a lot of time in planning and control; more than half of the CEO's calendar is consumed in planning and the COO and other top executives are even more engaged in our annual control cycle. This level of detail, along with the continuous repetition and iteration in the process is a distinguishing feature of Emerson's management process.*

THE OLD AND THE NEW

Since the nineteenth century, operating planning and budgeting cycles have been conducted annually, either starting January 1 and ending December 31 of the same year, or based upon the organization's fiscal year. This has been a tried and proven approach. Even the US federal government operates under this tried-and-true approach. It is considered a best practice to prepare an annual plan and supporting budget each year and stick to it. At IBM, managers were expected to manage their annual budget within plus or minus 5%. We believe that an annual budget is the right

* Knight, Performance without Compromise, 17.

approach for many organizations, but for a few leading-edge organizations, it may be time for a change.

> How can you manage a business in the twenty-first century using nineteenth-century planning and budgeting approaches?
>
> — **H. James Harrington**

At IBM we started our annual planning process in July and six months later it was completed. By late in December, we knew how many resources (people and money) we had in order to operate over the coming year. The problem was that many things had changed and the plans we generated so carefully in July were already obsolete. In this fast-changing business environment, we need to be able to quickly react to new opportunities as they arise. To take maximum advantage of the fast-changing environment, some organizations have abandoned the annual planning and budgeting cycle in favor of a rolling forecast and action plan that is used to update the rolling operating plan. To accommodate the older conservative proven approach and the new more risky rolling Operating Plan approach, we will be presenting both approaches in this chapter. Both of these approaches work, but we like the rolling Operating Plan approach better. We realize that many of you successful business leaders and chief financial officers will feel much more comfortable with the tried-and-true best practice of annual Operating Plans, but we ask you to at least consider that there is another way to control your organization.

PREPARING THE ANNUAL OPERATING PLAN

At least once a year a major part of managerial time is devoted to preparing the Annual Operating Plan. In big companies, like IBM, Xerox, Ford, and General Electric, that can take months, and for the state and federal government, it takes even longer. We have broken this process down into nine different activities as follows:

- **Activity 1:** Define Annual Operating Plan ground rules and assumptions.
- **Activity 2:** Prepare a proposed resource request.

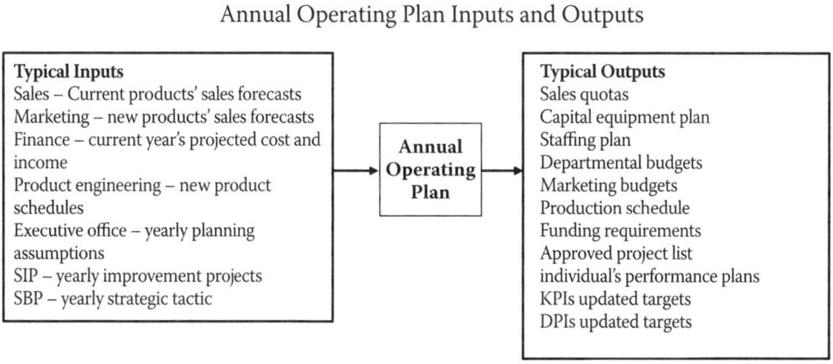

FIGURE 10.1
Annual Operating Plan typical inputs and outputs.

- **Activity 3:** Present proposed resource request to the Annual Operating Plan Review Committee.
- **Activity 4:** Present the resource request and projected incomes to the executive team.
- **Activity 5:** Revise the resource request based upon the executive team input.
- **Activity 6:** Obtain review and approval of the revised resource request from the executive team.
- **Activity 7:** Present the Annual Operating Plan to the board of directors and get its approval.
- **Activity 8:** Send approved related parts of the Annual Operating Plan to individual departments.
- **Activity 9:** Prepare individual Performance Plans for the employees.

The Annual Operating Plan takes a number of inputs, manipulates them, and produces a number of outputs (see Figure 10.1).

Activity 1: Define Annual Operating Plan Ground Rules and Assumptions

- **Planning ground rules:** Planning ground rules are the conditions, and the requirements on which the plan is based. The organization's current conditions are a key input to the planning ground rules. The requirements include setting targets for the organization's key performance indicators (KPIs).

- **Planning assumptions:** Planning assumptions are the conditions and inputs that the individual, who is doing the planning, will consider as part of his/her estimation. They are things that the planner believes will happen during the planning period. (Example: Productivity will increase by 10% or we will start shipping the new product on July 15.)
- **Key performance indicators (KPIs):** Key performance indicators are a set of measurements, usually no more than ten, which are used to evaluate the progress and performance of a total organization.

The first thing that needs to be done in a planning cycle is to define the things that can and will impact the planning results. We are calling these *planning ground rules* and *planning assumptions.* These are the inputs to the planning cycle that will impact the planning results.

The following are typical Annual Operating Plan guidelines:

- Sales forecast for product will be by month for the next fiscal year.
- The plan will be for the next fiscal year, in detail by month, and by quarter for the following two years.
- Estimates will be accurate to plus or minus 5% for the coming year and plus or minus 15% for the following two years.
- Information technology (IT) will be outsourced by May of next year.
- The average salary will be increased 2% per year.
- All planning will be bottom-up and based upon the processes within the unit.
- There will be a 10% increase in productivity in all departments.
- There will be no new hires. People who leave can be replaced with temporary employees for up to six months.
- Marketing will provide forecasted sales figures by month for all new products that are scheduled to be released during the next fiscal year.
- The plant will be shut down for three weeks to upgrade equipment and for vacations in August.
- Division 1, 2, and 8 will have a customer relationship management system installed that will require training all of the customer-interface personnel for approximately 80 hours.
- All employees will get a minimum of 40 hours of training per year.
- Item 1.7 from the Strategic Plan will start implementation on June 1. All other scheduled parts of the Strategic Plan will start as defined in the plan, with the exception of item 11.03; it has been put off for 2 years.

The following are typical Annual Operating Plan assumptions:

- The price of oil will go up by 25%.
- Our competition will start selling an advanced unit in the second quarter that will increase their share of the market.
- There will be an increased focus in the United States related to green activities.
- Interest rates will increase by 2.5%.
- Turnover rates will be approximately 8%.
- The stock market will not start to recover until the end of the third quarter.
- Raw material costs will decrease by 7%.
- The new MIP system will reduce inventory by 50%.
- The US Army will require more products to handle their increased involvement in the war on terrorism.

Assessing Your Current Position

This probably requires the most difficult soul-searching that the organization has done in a long time. It demands an honest appraisal of your company's present status, including each aspect of the organization that is deemed critical to its success. Consider things such as the following:

- Products
- Distribution channels
- Market share and influence
- Customer loyalty
- Technological innovations and advantages
- Management talent
- Employee capabilities
- Fulfillment capacity and equipment
- Competition
- Pricing

The role of each of these attributes in furthering our goals is not of specific concern to us just yet, for we haven't established the annual goals. Instead, the Annual Operating Plan calls for an honest appraisal of where the organization stands today. Using this information and comparing it with the soon-to-be-created annual goals allows us to see a gap emerging between where we are and where we want to be. It's that gap that much of our Business Plan deals with and closing that gap is the focus of the Annual Operating Plan.

Identification of Annual Performance Goals

From the beginning, you will work to create clearly defined goals for your Annual Operating Plan. These goals begin at the top of the company and cascade their way down in a "catchball" fashion. Some are longer range, twelve-month time horizons. Other subgoals needed to reach the larger targets are intermediate range, say, six months. Some are just three months out or even less. These are usually extremely critical things that must happen before work can proceed on the next step of a larger goal. These goals are largely driven by the Strategic Plan. Remember that the Strategic Plan and customer demand are the two major drivers of the Operating Plan.

Unlike those of some planning exercises you may have experienced, the catchball goals are brutally precise. They are the product of the organization's planners recognizing the specific targets the company must meet during the next twelve months if it is to prosper. It's easier for those who implement the plan to communicate precise goals rather than fuzzy, broad goals, and we want clarity. Since the annual time frame is short (12 months or more), there's no room for anyone wasting time doing something that doesn't specifically help the organization get where it wants to go. To this end, there are three questions we ask when establishing company goals:

1. Where do we want to go in terms of products, customers, profits, or return on investment?
2. What changes do we have to make in each department to get us there?
3. What needs to be done to accomplish the yearly objectives that are defined in the Strategic Plan?

Attainable Goals

It is important to recognize that there may be too large a gap between the goals you come up with and the organization's current position. There may come a time when you say, "There's no way we can accomplish all our goals this year." It takes a smart manager to recognize that the chasm is too wide to jump in a single leap. Instead, he or she must reassess the goals, and maybe scale them back a little or a lot. However, they must be realistic targets that everyone believes in. The outcome is greater commitment and higher probability of success than if you tried to ram unnecessarily burdensome goals down everyone's throat.

The trick to setting attainable goals is to strike a balance between targets that represent a realistic stretch and those that are so far out that they'll certainly cause disappointment and failure. On the one hand, identify something worth shooting for, and reward the team appropriately. On the other hand, people view impossible goals as negative even when they are sweetened with better rewards.

Once you've identified your current position and arrived at the annual targets for your company, it's time to figure out how to bridge the gap between the two. The technicians call this process *reverse engineering*. In essence, we're taking the end result and working backward to determine the steps we need to get there.

Typically, one organization is assigned to coordinate the Annual Operating Plan. Often coordinating the preparation of the Annual Operating Plan will be assigned to the strategic planning officer's department and the Finance Department working together as a team. The strategic planning officer's department is assigned because the Operating Plan must be based on and in line with the Strategic Plan, and the Finance Department is assigned because the organization's budget is such an important output from this cycle. The manager responsible for coordinating the Annual Operating Plan will assemble an Annual Operating Plan Review Committee. At a minimum, one person from the Strategic Planning Office, Finance, and Human Relations should be on this committee. Often this committee consists of representatives from Product Development, Sales and Marketing, Quality Assurance, Industrial Engineering, Human Relations, Finance, and Manufacturing Engineering. The Annual Operating Plan Review Committee will be responsible for

- Preparing the ground rules package
- Preparing the Annual Operating Plan cycle schedule
- Defining input requirements and format
- Reviewing the individual department's Annual Operating Plan to be sure that it is aligned with the guidelines and the Strategic Plan, and that the resources requested are justified
- Determine if the goals and objectives for the individual departments are in line with and supportive of the improvements that need to be made

- Review the risks related to the individual department's Annual Operating Plan and budget to be sure they have been addressed
- Make recommendations to the executive team to accept or reject the individual budgets

Typical things that are included in the ground rules package include

- The schedule for preparing the Annual Operating Plan
- The assumptions and ground rules on which the plan is based
- Production schedules
- Sales forecasts
- Approved projects
- Strategic Business Plan activities scheduled for the next fiscal year
- Strategic Improvement Plan activities scheduled for the next fiscal year
- End of current year projected cost and revenue for each department
- Required forms for presenting to the Annual Operating Plan Review Committee
- Forms that are used to calculate workload and prepare budgets
- New product release schedules
- Other

Activity 2: Prepare Proposed Resource Request

- **Department performance indicators (DPIs):** A set of measurements, usually no more than eight, that are used to measure the performance of a department.
- **Individual performance indicators (IPIs):** A set of measurements that are used to measure the performance of the natural work teams throughout the organization, the individual managers, and the employees. These are key inputs to their performance plan and evaluation. Often they are included in the individual's job description.
- **Natural work team (NWT):** A group of people who are assigned to work together and report to the same manager or supervisor. Natural Work Teams exist at all levels of the organization.
- **Annual business planning process:** The combination of reviewing and updating the Strategic Plan and preparing the Annual Operating Plan.

Goals at the NWT level are defined by identifying the major processes that the NWT performs and then establishing effectiveness and efficiency requirements for each of them. A methodology called *area activity analysis* is used to establish these requirements. The effectiveness requirements are used to define the minimum quality requirements and the efficiency requirements are used to define the minimum cost and the maximum number of people that are required to perform the activities that make up the process. Each year a 5% to 15% improvement over last year's performance should be required for each measurement.

The first activity in defining resource requirements is to establish annual improvement goals for all of the key processes measurements. Why establish annual goals? Many organizations focus on making money today instead of investing time in the annual business planning process. Those are the people who lack a definite direction and the ability to stay on course. With changing technology, changing customer demands, and increasing sophistication, marching in place is business suicide.

These yearly improvement goals are the targets for change and transition that your organization must reach over the planning horizon—for our purposes, the next twelve months. These goals, whenever possible, should be driven by the goals defined in the Strategic Plan. Company-wide goals established in the Annual Operating Plan move the company into the position where it needs to be.

The following are typical department performance goals for a financial department:

- Get additional funds of $1,000,000.
- Purchase and take delivery of 5 new knitting machines.
- Expand service territory into China by Dec. 2010.
- Generate added sales volume of $2,000,000.
- Help attain the overall profit or cost reduction objectives of $3,000,000.

Failure to reach any one of these annual department goals could jeopardize reaching the company's overall target. Additionally, within every department, it's easy to identify exactly what that department must do to further the company's cause.

All too often, managers take last year's actual cost and add the projected *cost of living* increases and the cost of doing new projects to come up with their new budgets. The advantage to this approach is that it is very fast and easy to complete. The disadvantage is that after just a few years the budget becomes inflated and the department begins to use more resources than are required. The proper way to define the resource requirements is to define the processes in which each department is involved and measure the time required to process an item through the individual processes. Then, by measuring the number of items processed in the past year, the manager can estimate the number of items that will be processed in the coming year. This approach provides you with the justification for your HR requirements, which is a key part of the inputs to the next fiscal year's budget.

There are five parts to the resource requirements package. They are

- Department training plan
- Department capital equipment requirements
- Department human resource requirements
- Department budget
- DPI targets

Each department at all levels in the organization is required to prepare a resource request package that is reviewed and approved by the next higher-level manager. Once this manager has signed off on the department's resource request package, it is ready to be presented to the Annual Operating Plan Review Committee.

Activity 3: Present Proposed Resource Request to the Annual Operating Plan Review Committee

This is a very important meeting. Although the Annual Operating Plan Review Committee can't accept or reject the resource request, their recommendations are usually accepted by the executive team and without their support for the resource requests, most executive teams will not even schedule a meeting with the related manager. As a result, you need to be very well prepared for these meetings. Usually, the Annual Operating Plan Review Committee will already have reviewed the data used to prepare last year's Annual Operating Plan for the department. They will have reviewed the department's present budget versus actual and studied the status of the DPIs. They will also know how many people the department has on board and their skill levels. The makeup of this committee also allows them to probe very deeply into the assumptions that the individual departments make as they prepare their resource requirements.

We have sat in on many of these meetings, both as presenters and as reviewers. There have been very few of these reviews where the Annual Operating Plan Review Committee hasn't found some problems or made some suggestions that require revision of the department's resource requests and presentation of the revised request to the Annual Operating Plan Review Committee before it recommends the request to the executive team. Often the resource request is rejected because the DPI targets are not aggressive enough.

Activity 4: Present the Resource Request and Projected Incomes to the Executive Team

Typically, the executive team is interested in reviewing a bigger picture than the Annual Operating Plan Review Committee. They will often wait until all of the departments that make up an individual function have been approved by the Annual Operating Plan Review Committee and then schedule a meeting where the total resource requirements for the function are reviewed, first in total and then for specific departments in which they are interested. They usually use the present year's resource consumption as a benchmark to highlight problems. Functions that are generating revenue will also present their revenue projections. There is some pressure at this meeting to reduce resource requirements if they aren't in line with the present year's requirements. The real pressure from the executive team comes when the inputs from the total organization are combined and a profit and loss analysis can be made. The problem with most organizations is that at the department level, the manager puts in a safety factor along with some things that he/she would like but does not need. The functional manager adds in another safety factor and so does the chief financial officer.

On the other side of the equation, the Sales and Marketing team play it conservative when it comes to revenue generation. As a result, the executive team often rejects the total Annual Operating Plan to bring the next fiscal year performance in line with what they believe can be accomplished. This often results in a 10–20% cut in resources and dropping or slipping some of the proposed projects that don't have an extremely high return on investment. Often some of the projects that are slipped are projects that were part of the Strategic Business Plan and/or the Strategic Improvement Plan. This results in the need to revise both of these plans.

Activity 5: Revise the Resource Request Based upon the Executive Team Inputs

The individual departments will then go back and revise their resource requests to stay in line with executive team inputs.

Often the functional manager will decide how much the individual departments will need to reduce their resource requests. Once all of the functional total resource requests are in line with what the executive team requires, another meeting is scheduled with the executive team.

Activity 6: Review and Approve the Revised Resource Request by the Executive Team

The second meeting with the executive team is usually much shorter and focuses on determining that all of the key strategies are still supported, that the total resource request is in line with the executive team direction, and the DPI targets are aggressive enough. It is very important to everyone that the stockholders make a fair return on their investment. A publicly owned organization will also focus on insuring that the Annual Operating Plan has a positive impact on its stock prices.

Activity 7: Present the Annual Operating Plan to the Board of Directors and Get Its Approval

One of the major board of directors' committees is the Finance Committee, which will review the Annual Operating Plan, focusing primarily on the financial implications, and then focusing on targeted change projects and the KPI targets for the next fiscal year. This committee knows that the board of directors is held accountable for the organization's performance and its ethics in the same way as the executive team. As a result, they will do a very thorough analysis.

Remember, the board of directors represents the investors who have given their hard-earned money to the organization so that it can operate. Much of this money comes from retirement funds and small investors who are depending upon it to supplement their small social security check that comes in monthly. (For example, I worked for IBM for 40 years and all that time I invested 15% of my pay into the IBM Employee Stock Purchase Plan, which was a big part of my retirement fund. When IBM cut their dividends, it had a big impact on my living standards.) When stock prices go down, this often means that the children's college fund is cut so badly that they can no longer afford to go to college. As a result, the Board of Directors Finance Committee wants to see an Annual Operating Plan that produces an extremely high return on investment and increases stock prices. This committee will want to cut costs, increase sales, and put projects that don't have a short-term impact on profits on hold. As a result, the Annual Operating Plan is often revised based upon the board of directors' Finance Committee recommendations.

Once the board of directors has approved the Annual Operating Plan, it can be finalized.

Activity 8: Send Approved Related Parts of the Annual Operating Plan to Individual Departments

When the Annual Operating Plan is finalized, it is broken down to the department level and each department receives information that is relevant to their operation. For example:

- Budgets broken down by month for the next fiscal year
- Human resource allocation numbers by month
- Approved capital equipment budget and a list of approved capital equipment
- Approved training allocations
- Approved DPIs
- List of approved projects
- Production schedules
- New product schedules
- Pricing information for each product

Basically, this is the key information every manager needs to run his/her department.

Activity 9: Prepare Individual Performance Plans for the Employees

It is easy to see that activities 1 through 8, with all the recycles, can take a long time to complete in a mid-sized or large organization. It often takes 3–7 months to go through the entire cycle. As a result, the information required to do effective performance planning for the managers and employees usually becomes available at or near the beginning of the next fiscal year. It is very important that the performance plan for your employees and managers is closely tied in with the Annual Operating Plan commitments made by that department. Tying the individual performance plans to the department commitments (DPIs) allows the manager to develop for each person a set of individual performance indicators (IPIs) that can be measured as part of their performance evaluation. If the Annual Operating Plan calls for a 10% improvement in productivity, this 10% improvement needs to be reflected down into the IPIs that make up the employees' performance plan and used to evaluate their performance throughout the year.

HOW TO DEVELOP THE ANNUAL OPERATING PLAN FOR A SMALL ORGANIZATION

A small organization needs an Annual Operating Plan but often can't afford the effort and time required to do an Annual Operating Plan as it is done in a midsize or larger organization. However, a small organization's top management, with some help from an experienced consultant, can build the Annual Operating Plan quickly. Five activities are usually required:

- **Activity 1, Launch:** This involves 2 to 5 days of hard work defining the operational goal, the key areas of commitment, and the start of detailed commitment planning.
- **Activity 2, Build commitment to the annual operating plan through a schedule of events:** This activity includes a period of 3 to 6 weeks during which there are several days of detailed commitment planning. This usually requires frequent checks with the consultant by the executive team.
- **Activity 3, Finalize the plan:** This involves 2 to 5 days of hard work. The result is the final operational plan including all documentation and an execution dashboard.
- **Activity 4, Checkpoints:** This activity occurs three months after finalization of the plan. There is a rigorous review of progress and the decisions that were made and is repeated each quarter.
- **Activity 5, Celebrate:** If the goals are achieved within deadline and budget, it is time to give out rewards and celebrate the success of the Annual Operating Plan.

THREE CHARACTERISTICS OF AN EFFECTIVE ANNUAL OPERATING PLAN

There are three important attributes to a good Annual Operating Plan.*

1. There is an appropriate level of detail; enough to guide the work, but not so much that it becomes overwhelming, confusing, or unnecessarily constrains creativity
2. There is a format that allows for periodic reports on progress toward the specific goals and objectives.
3. There is structure that aligns the vision and values with the Strategic Plan;† the goal statements for the Strategic Plan and the Annual Operating Plan are integrated and can be considered as one and the same. The objective statements for the Strategic Plan and the Annual Operating Plan will be different; however, the Annual Operating Plan objectives will be supportive of the Strategic Plan objectives as they cover a one-year period, whereas the Strategic Plan objectives cover a 5- to 10-year period.

Just as monthly financial statements often present a budget for revenues and expenses and then report actual figures for a given time period, so should the Annual Operating Plan allow for the same type of comparison: the plan declares the "budgeted" work in terms of goals and objectives for each program area and management function, and reports the actual progress on a monthly or perhaps quarterly basis. This *budget-to-actual* report gives a clear reading on how the "trip" is going.

The MyWorkTools.com‡ spreadsheet is a template for developing an Annual Operating Plan for the entire organization.

---

* Knight, Performance without Compromise, 18–19.

† Vision and values are the obvious places to start. You need to know the current and future vision and goals of the organization and find out how closely operations in each part of the organization align to the overall vision and goals. Next, communicate any major differences between the goals of a business area and the overall goals of the organization. You also need to know what worries senior management—what are the current key problems? Finally, check if everyone in your learning and development team knows the organization's vision, goals, and drivers.

‡ MyWorkTools.com is the world's first knowledge exchange focused exclusively on digital business tools. They purport to have the largest selection of proven, quality tools created by experts, consultants, and professors. They do not create the tools themselves, but provide consistent tool descriptions, pricing, installation, and quality control. All tools have a money back guarantee.

- **Background:** Developing an integrated Annual Operating Plan is a requirement for successfully establishing and managing any business. This tool integrates all of the key components—income statements, sales forecasts by product, operating expenses, capital expenses, depreciation, headcount by department, and much more. All of the departments are set up, but can be changed to fit your organization.
- **Benefits:** Using these tools, you will be able to increase business efficiency—refine sales forecasting, improve expense management, enhance headcount management, approve major operating expense or capital expense programs—and understand how changes in these items affect your financial statements. The Annual Operating Plan will enable you to financially plan each element of your business and facilitate quick decision-making processes.*
- **Time ROI:** It takes on average of at least 100 hours to develop a detailed, integrated Annual Operating Plan for the entire small organization. By using MyWorkTools.com, you will save approximately 80 hours since all you have to do is input organization-specific information, which takes about 20 hours to complete. This information is then propagated throughout the Annual Operating Plan for all key departments (see Figure 10.2).

ROLLING OPERATING PLAN

In some organizations the Annual Operating Plan process is out and a new Rolling Operating Plan serves as the basis for day-to-day operations. Unilever, for example, stopped doing annual budgeting in 2010. The Annual Operating Plan has been replaced with event-driven planning, flexible budgets, and rolling forecasts. Neil Vorchheimer, senior vice president of finance for Unilever, stated: "We used to have what we called the annual plan, and we'd spend six months of the previous year putting

* Make sure you have quick decision-making processes. You have to be quick on your feet to keep up with the changes within your organization. To be successful, you need to (a) ensure that your annual operations strategy responds quickly to changes in the organization's overall strategy, (b) develop champions who are committed to supporting your learning initiatives, and (c) reduce the number of touch points or points in which you need sign-off.

DELTA **Operational Sales Plan -Example**

Progress vs Monthly Milestones												Plan					Goal		
												Commitments				Obstacles	NC	CSF	Operational Goal
J	F	M	A	M	J	J	A	S	O	N	D	Budget	Due	Owner	Deliverable				
													July	FB	1.11 ACA for all key accounts done	Account constraints are unknown	1.1 T-Funnel of 200M EU	1) Break-throug in Key Accounts	120 M EU in 2008
													Feb	KS	1.12 OpMaps reviewed at account meetings	OppMap is not reviewed with accounts			
													Jan	KS	1.21 All projects > 10M have managed win plans	Only few opportunities have win plans	1.2 >50% Win rate in T-Projects		
													March	AF	1.22 W/L reviews for all projects >10M	We do not learn from success or failure			
												125 k$ CRM	Jan	NE	2.11 C-reach reviewed at monthly management meetings	C-Reach goes untracked	2.1 C-Reach >75%	2) Strong growth in base business	
													Jan	BR	2.12 Owner selected and defined	Ownership is unclear			
												90 k$ CRM	Jan	DL	2.21 StR reviewed at monthly management meetings	Ownership and goal are undefined	2.2 >30% Straight rebuy		
													Jan	DL	2.22 Improvement plan reviewed at monthly management meetings	Improvement plan is not defined, confirmed			
													Jan	BR	3.11 6 mths PD Cycle is established as one of "top ten goals"	Goal is not key company performance standard	3.1 <6 months for new products	3) Fast product develop-ment	
												200k$ Con-sult	March	BR	3.21 Project progress reviewed weekly (Friday afternoon)	Project methodology is unknown			

FIGURE 10.2
Delta's operational sales plan part of the Annual Operating Plan.

it together. As soon as the budget was approved it was out of date. So we decided to do away with it."*

Unilever is not the only organization that has put aside the Annual Operating Plan in favor of a Rolling Operating Plan. Organizations like Norton Lilly International and American Central Investments have also embraced the same approach because the business environment is changing so rapidly. The Annual Operating Plan is too rigid to provide the flexibility needed to make quick decisions that are often imperative in today's business world. The chief financial officer of Clements International, A. J. Chopra, points out: "You lose sight of the fact that the strategic view is never revisited during the year."†

* Neal Vorchheimer, "Good-Bye Annual Budget. Hello Rolling Forecast" May 3, 2011, *CPA Trendlines.com*, http://cpatrendlines.com/2011/05/03/good-bye-annual-budget-hello-rolling-forecast (accessed December 5, 2011).

† Choptra, A. J., *Managing the People Side of Innovation: 8 Rules for Engaging Minds and Hearts.* (West Hartford, CT: Kumerian Press, 1999). Also, Banham, R., "Let It Roll: Why more companies are abandoning budgets in favor of rolling forecasts," *CFO Magazine*, May 1, 2011, http://www.cfo.com/article.cfm/14570220 (accessed 5 December 2011).

The concept of a Rolling Operating Plan makes the Strategic Plan and the organization's budget living documents that are not cast in concrete; they are constantly updated to reflect the changing environment, changing customer demands, taking advantage of unseen opportunities, slips and advancements in approved projects, and newly approved projects. It greatly increases the CFO's and the strategic planning officer's involvement in the day-to-day running of the business. Steve Player, program director at the Beyond Budgeting Roundtable, stated: "In the old days, the CFO sat in the back of the ship recording what happened. Now, the CFO stands on the bridge looking forward and adjusting for variables."*

Typically in a Rolling Operating Plan, the Strategic Plan and the budgets are adjusted and updated on a quarterly basis, although some organizations do it monthly. These changes reflect the following:

- Cost savings from projects; for example, Six Sigma projects or installing a customer relations management (CRM) system
- Changes in demand forecast
- Changes in project schedules
- New product releases
- Changes in customer requirements
- New government regulations
- Organizational structural changes
- Newly approved unscheduled projects
- Other

Changes that affect cost, schedule, budget, and planning are accumulated by the finance or strategic planning office as they are approved by the executive team and/or the board of directors. Then the finance office prepares the revised budget and sends it out to the individual departments for their review and approval. Often, causal models are developed and computerized that identify the impact that individual changes will have on the various departments throughout the organization. The Strategic Planning Department uses the same approach to update the Strategic Plan based on the status report from the individual departments and the management reviews. Strict ground rules are defined about the amount of money that is added or removed from the budget before the board of directors or the

* Steve Player, "Why You Must Move Traditional Budgets to Achieve Lean Success," *Beyond Budget Roundtable*, Lean Accounting Summit Workshop presentation, http://www.bbrtna.org/resourcelib.html.

board of directors Finance Committee need to approve the change. All major changes to the Strategic Plan are reviewed at the board of directors meeting to ensure good corporate governance.

Although Rolling Operating Plans eliminate much of the planning and budgeting work from the individual department, they use the individual performance indicators that are generated by the departments as a basis for the finance department's bottom-up budget analysis. As Tony Tripeny, corporate controller at Corning, states: "We do rolling forecasts that we update each month to address what we think will happen for the rest of the quarter. Based upon this analysis, we will go to the business unit and say, 'What are you going to do differently? What actions are you going to take, and how is that different from what we had assumed with the budget?'"*

THE COMPROMISE

Many organizations do not feel comfortable relying solely on a Rolling Operating Plan. They like the more detailed involvement of the individual department in the Annual Operating Plan process. During these cycles, specific objectives are defined for each department and the revised IPIs are documented. Many controllers feel that the annual budget is still of value as it provides a starting point, but they have added a quarterly review and update as part of the process to be able to react to many changes that occur in the business. The strategic planning office also likes the idea of getting a once-per-year detailed review of the Strategic Plan that is closely tied in with the individual department's resource allocations.

There is universal agreement that there is a big exposure to relying on† an Annual Operating Plan. A survey conducted by The Hackett Group (*2011 Key Issues Study*), reported that the uncertainty in the following, as measured from before the 2008 financial crisis to the 2011 level increased as follows:

- Demand volatility: up 67%
- Volatility of market prices of our outputs: up 67%

---

* Tony Tripeny, telephone interview 4/11/2010, with author Frank Voehl.

† The Hackett Group 2011 Key Issues Study published in "Let It Roll: Why more companies are abandoning budgets in favor of rolling forecasts," *CFO Magazine*, May 1, 2011, http://www.cfo.com/article.cfm/14570220.

- Overall level of volatility of business conditions: up 60%
- Volatility of input prices (raw materials, resources): up 52%
- Exchange-rate volatility: up 49%
- Costs of and/or access to financing: up 24%
- Available talent to meet the workforce head-count plan: up 7%

American Century Corporation uses a six-quarter, forward-looking view of its business at the end of each quarter. Jon Zindel, CFO of American Century Investments stated: "We then determine whether to add new projects or adjust our plans based on discussions of trends in the industry, regulation changes coming at us, or competitive or economic issues. Resources are allocated based upon actual demand, which is constantly monitored. We are a highly adaptive firm now. We're no longer making a bunch of assumptions in a budget about what will happen, and then being wrong by some factor."*

SUMMARY

On one hand is the Strategic Plan and on the other is the Annual Operating Plan. Surprisingly, while there is much attention placed on the Strategic Plan, the Annual Operating Plan in startups and small businesses often receives very little attention. For startups and small business owners, sometimes it has been falsely believed that having a detailed Annual Operating Plan is only an option and was not required. Many entrepreneurs see that kind of planning as reserved for larger companies where there are more personnel. That is not the case any longer. It's time to get down to the nitty-gritty†—the day-to-day operations of your business, including location and hours of operation, manufacturing schedule, and the internal operations and equipment required to produce your product or service. Your Annual Operating Plan demonstrates that you've thought

* Russ Banham, "Let It Roll: Why more companies are abandoning budgets in favor of rolling forecasts" *CFO Magazine*, May 1, 2011, http://www.cfo.com/article.cfm/14570220.

† Explain where your business is located or will be located, and include the size and type of space, including square footage and whether it's an office, manufacturing facility, retail space, or warehouse. Describe any advantages (or disadvantages) to your location, and if relocation is possible, explain when it will be necessary and why. (Include a rough layout of the property in the appendices.)

through—or you're already using—a viable approach to creating your product or service and getting it delivered.*

An Annual Operating Plan is just as relevant to small startups as it is to large corporations. It is basically a detailed list of events and responsibilities that need to be done correctly in order to achieve the goals and objectives formulated in the Strategic Plan. It is important for any organization, large or small, to have an Annual Operating Plan that corresponds to a fiscal year for each department. It is also essential that the Annual Operating Plan take full advantage of the business intelligence opportunities that the planning process often provides managers and owners who are savvy enough to take advantage of the opportunities.†

Having an Annual Operating Plan ensures that all employees know their responsibilities and are able to coordinate their own efforts with other employees to accomplish a task. In addition, it allows managers to track the progress of the Annual Operating Plan and whether or not it will finish on time. This is important, because if they do need funding ahead of time, they are able to request it ahead of time.

One way to think of the importance of an Annual Operating Plan is to think of this scenario: You are going on a long road trip to get to a certain destination for a business conference. It is a trip that will require more than a day's travel. Therefore, planning is definitely needed to determine which roads are the best to take, when to take rests for food and relaxation, and when to refuel. Moreover, you are on this trip with your colleagues who have their own needs and their own responsibilities. Thus, there are several scenarios to consider. Is there a need to rotate drivers? Maybe if you can switch drivers, there is a possibility of just driving straight through

* List and describe all the equipment you require, from office to vehicles to manufacturing machinery. Explain what each does and how they work together. Include the cost to purchase, lease, and maintain the equipment. Include details on all your employees, including full-time, part-time, and outsourced. Create a table that lists the number of employees, hours worked, and pay, organized by overall job function. Include job descriptions and skill sets. Explain the production, manufacturing, or service process from beginning to end. List your key suppliers and how you pay for and receive materials. Where do you store finished products, and what does this cost? How do you distribute finished products? How long does the entire process take? How do you track inventory?

† To fully exploit operations planning for sustainable competitive advantage and to deliver the results executives expect, planning and data management initiatives must be tightly linked to business goals and objectives. You must align technology with business objectives and engage key stakeholders in this effort. Such alignment is far from easy. It is a challenging task that requires the right set of skills and experience in business processes and the right data management initiatives in place that work effectively."

without staying one night at a hotel. Which person is responsible for food and the pocket money for the trip?

As you can see, having an Annual Operating Plan would be very helpful in detailing how to get the entire group to work cohesively as a unit in the same direction. To craft an effective Annual Operating Plan, there are some vital attributes that are usually evident. First, an Annual Operating Plan should be constructed in a way that works synergistically with the overall Strategic Plan. While the immediate objectives will differ, the end goals for both plans should be the same. Second, an Annual Operating Plan should more or less serve as a guideline. The right amount of hiring and retention of key people having extraordinary talent, technical expertise, self-motivation, team spirit, and company loyalty will be crucial to your success and growth, and are an integral part of the Annual Operating Plan.

Therefore, shared accountability and responsibility will be a hallmark of your Organizational Master Plan personnel policies (which have been proven by legendary and very successful companies). Generous stock options may sometimes be offered as a sign-on incentive, and especially as a merit bonus for outstanding individual and team achievements. Moreover, you will need to foster work performance excellence, by mentoring new employees and by encouraging open and objective technical discussions and critiques.

You need to remember that the Strategic Plan can only carry your organization so far, and upon completion, an Annual Operating Plan for each upcoming year (or two) must be prepared. An Annual Operating Plan is a schedule of events and responsibilities that details the actions to be taken in order to accomplish the goals and objectives laid out in the Strategic Plan for that point in time. The Annual Operating Plan ensures that everyone knows what needs to be done, coordinates their efforts to get it done, and can keep close track of whether and how it got done.

> It takes a short range plan to be sure that the long range plan and objectives are accomplished.
>
> **—H. James Harrington**

EPILOGUE

Charles Knight, chairman emeritus of Emerson Electric, one of Americas most admired companies offers the following:

Throughout these challenging times, we stressed three major points. First, we wanted to get restructuring done as quickly as possible. Second, we wanted to focus the business investments annually, so that when the economy rebounded, we could accelerate sales and earnings growth ahead of the competition. And third, we wanted to significantly improve our asset utilization as measured by free cash flow and return on total capital. A key part of the annual execution plan involved communicating to the entire organization the issues we faced, the actions to be taken, our near-term and long-term goals, and our expectations of what the company would look like when we cam through this restructuring period. Consequently, we began a live quarterly global Webcast to lay out the issues and actions. Second, we laid out an extensive Strategic Plan and communicated it at the yearly planning conference. Moreover, each year we have followed up with employees and the Board of Directors on where we stand relative to these goals and objectives.*

—H. James Harrington

* Knight, Performance without Compromise, 225.

11

Organizational Master Plan Summary

Why is it that university professors get paid so little in comparison to CEOs? It is because it doesn't require much ability to know what to do. It is making it happen where the real value is added.

—H. James Harrington

INTRODUCTION

Now that you have an Organizational Master Plan all together, the real stamina of the organization's ability is put to the test. It is at the point in the process where words on paper are magically transformed into dollars. It is the point where promises are changed into action. It is the time when loose talk is turned into tangible results. It is easy to talk about the work you are going to do tomorrow and much harder to get that work done tomorrow.

When Dr. Harrington was working for IBM in the 1980s, IBM hired the top-rated consultant with the best reputation to help with one location's improvement plan. The consultant assigned two men to work full-time with the IBM team at the location. The three of them worked a full year, which cost IBM well over a million dollars. They produced a set of documents that included goals, operating procedures, check lists, and so on. The IBM team devoted hundreds of IBM man-days to work with the consultants, which cost IBM more than the consultants' bill. After a year, the three consultants went away and the documents they prepared sat on one man's desk until he was reassigned and then they went into storage. All that work was wasted, not because it was a bad plan—for it was a very good plan—and not because it was poorly prepared or because it lacked support for the planning process. No—it was wasted because it never was accepted as part of the Annual Operating Plan. The follow-on money that was budgeted for it was siphoned off into other projects, much of it to fight fires that the plan would never have allowed to occur.

Somehow it is always easier to select a route that you know will get you where you want to go rather than trying a new route that someone else tells you will be faster and better. The challenge is turning plans into action; that is, having the resolution to set your rudder in one direction and keep it there. In many organizations it is almost as though the executive thought pattern is: "Oh, now the planning is done. Now we can get back to the real world."

The secret to implementing the Organizational Master Plan is the effective use of the Six Ts: talk, training, time, tools, teamwork, traceability.

Talk

It is essential that we keep the mission, vision, objectives, goals, and strategy in the forefront of everyone's mind throughout the entire organization. Everyone should not only know these key items, but understand their role in making them happen. All the managers need to walk the talk. The organization can't afford to have a single manager that is a black hole. A black hole in space is a whirlpool-like object that sucks all types of matter into it, but nothing comes out the other end. We have too many managers who are black holes, where information pores into them but doesn't move on through to the people who report to them, or if it does, it is distorted to the point that it is not effective. I have literally heard managers say something like this to their employees: "Would you believe it? Top management wants us to be more participative."

Our talk needs to be transformed into action. Our words need to change our behavior. We need to develop a completely new communication system that is designed to support the Organizational Master Plan.

Training

We need to provide our people with training to survive in the new culture. We need to provide them with training that allows them to effectively use the new technologies. We need provide them with training that keeps them challenged with new ideas. Most businesses are moving so fast that without the additional training, our people become obsolete very rapidly and as a result, they will be unable to cope with the changes that are required in the new culture.

Time

New ways require time to learn, time to adjust to the new procedure, and time to implement the new processes. Too many Organizational Master Plans are unsuccessful because the time required to implement them is not included in the budget. Managers rationalize that the additional work can just be absorbed by the employees because management believes their people are not really busy at the present time. The truth of the matter is, from the employee's standpoint, if you didn't budget for it, it really isn't worthwhile doing.

Tools

Our technologies are moving so fast that our tools, which used to last for 10 years, are becoming obsolete in 1 to 2 years. We have seen Organizational Master Plans that failed, not because of the plan, but because their competition had upgraded their hardware and software, allowing them to react faster and more accurately to customer requirements, while our organization was struggling with slower computer systems. We can't afford to handicap our employees with tools that are worn, broken, or obsolete.

Teamwork

We are living in an environment where the total organization needs to be aligned with the Organizational Master Plan. This means that the functional smokestacks that used to exist must be torn down. All our handoffs need to be much smoother. We used to think that teams were the answer, but today with things changing so fast, the team concept has given way to teamwork. People come together for a quick meeting to solve a problem, the team lasts for only one hour, and then it is disbanded.

More and more people are focused on teamwork among individuals who are concerned about their customers (internal and external) and suppliers (internal and external).

With the teamwork concept, everyone has time to stop and help someone who needs their help. This requires that everyone has an understanding of the Organizational Master Plan so they know how they interact with other organizations that are in the process of implementing it.

Traceability

Traceability is the key to keeping everyone informed about what they need to do to support the Organizational Master Plan and follow up on assignments to be sure they are completed on schedule. It puts the responsibility for notifying the affected people, whenever there is a slip on schedule, on the shoulders of the person who causes the slip. It requires mitigation plans to be developed so that this type of slippage will not occur again.

It is not enough to prepare a plan, define what is going to be done, fund the plan, and think that it is going to happen. Often it doesn't work that way. There are just too many day-to-day fires that get in the way of doing the things that prevent the fires from occurring. It is for this reason that we suggest that the entire plan for the coming 12 months be developed right down to the individual work breakdown structure, and be loaded into a computer program. With good project tracking software, anyone who has an assigned activity that is scheduled to start, is active, is scheduled to be completed, or is overdue, will be sent an update at least once a month. The individual assigned to each activity item can update it. Overdue items are updated on a weekly basis. The following information should be included on the report:

- Name of the assigned person
- Description of the assigned activity
- Original start and end target dates
- Latest start and end date targets
- List of status updates that includes the reason for any slippage

We like to have an escalation feature in the program where any action item that is two weeks overdue is automatically escalated and assigned to the manager of the person who is responsible for doing the task. This next-level manager is then responsible for investigating the situation and recording what action is being taken to get the activity back on schedule or to at least minimize the slippage. All schedule slippages require two levels of management approval.

COMMUNICATING THE ORGANIZATIONAL MASTER PLAN

The Organizational Master Plan isn't the executive's plan, it is not his plan, or her plan—it is our plan.

—H. James Harrington

Now that you have completed developing the Organizational Master Plan, it is time to communicate the plan. Some of the biggest stumbling blocks to all well-laid Strategic Plans are communications, time, and money. As with every business, budgets are never big enough to do everything you want to do. Prioritize and then communicate key goals by asking: Do implementing the goals make financial sense? Do we have the human resources to achieve our plan? How will the goals be communicated?

Note that certain groups of stakeholders might get complete copies of the plan, including appendices, while other groups (usually outside of the organization) might receive only the body of the plan without its appendices. The following is a sequence of communication timelines that should be scheduled, once the plan is in place:

1. Every board member and member of management should get a copy of the plan (Month 1).
2. Consider distributing all (or highlights from) the plan to everyone in the organization. It's amazing how even the newest staff member gains quick context, appreciation, and meaning from a review of the Organizational Master Plan (Month 2).
3. Post your mission, vision, and value statements on the walls of your main offices, within every manager's office, and on all the organization's bulletin boards. Consider giving each employee a card with the statements (or highlights from them) on the card (Month 2).
4. Publish portions of your plan in your regular newsletter, and advertising and marketing materials (brochures, ads, etc.) (Month 2).
5. Train board members and employees on portions of the plan during orientations (Month 3).
6. Include portions of the plan in policies and procedures, including the employee manual (Months 3–4).
7. Consider copies of the plan for major stakeholders, for example, funders and investors, trade associations, potential collaborators, vendors/suppliers, and so on (Month 4).

> Most managers manage for yesterday's conditions, because yesterday is where they got their experience and had their successes. But management is about tomorrow, not yesterday. Tomorrow concerns what should be done, not what has been done.
>
> **—Theodore Levitt, Harvard Business School**

The Organizational Master Planning process applies to the public and private sectors equally as well. The national government is now requiring it from all units. Figure 11.1 is a typical example of how a community transforms goals into action.

DEVELOP THE FORMAT, CONTENT, AND PLAN STRUCTURE

In Chapters 6, 7, and 8, management set performance goals that were measurable. In Chapters 9 and 10, additional performance goals, targets, and checkpoints were added by lower-level management and the employees. Put these measurements and targets on a scorecard (in Excel), which acts as an instrument panel guiding your organization toward achieving your vision. Next, develop a timeline for each measure. With the scorecard, you can actively track your progress on a monthly basis.

A leader devoted to the successful implementation of the strategy and plan is key. The plan needs to be supported with people, money, time, systems, and above all, communication. Communicate the plan to everyone in your organization. Hold a monthly or quarterly Organizational Master Plan Review meeting to report on the progress toward achieving the goals and associated actions. Build a Strategic Plan with very detailed activities spelled out for the next 12 months or so in the Annual Operating Plan. Don't forget to take corrective actions when needed and adapt as the environment changes.

The authors feel that it's wise to distribute the detailed timelines and copies of the plan to major stakeholders (investors/funders, trade associations, etc.). Therefore, you should organize the format of the plan such that the body of the plan can be sent outside of the organization and the appendices can include the more confidential and detail-oriented documents—documents that also tend to change a lot. The format of the plan should fit the culture and preferences of the organization. A typical planning document would include the following (Figure 11.1):

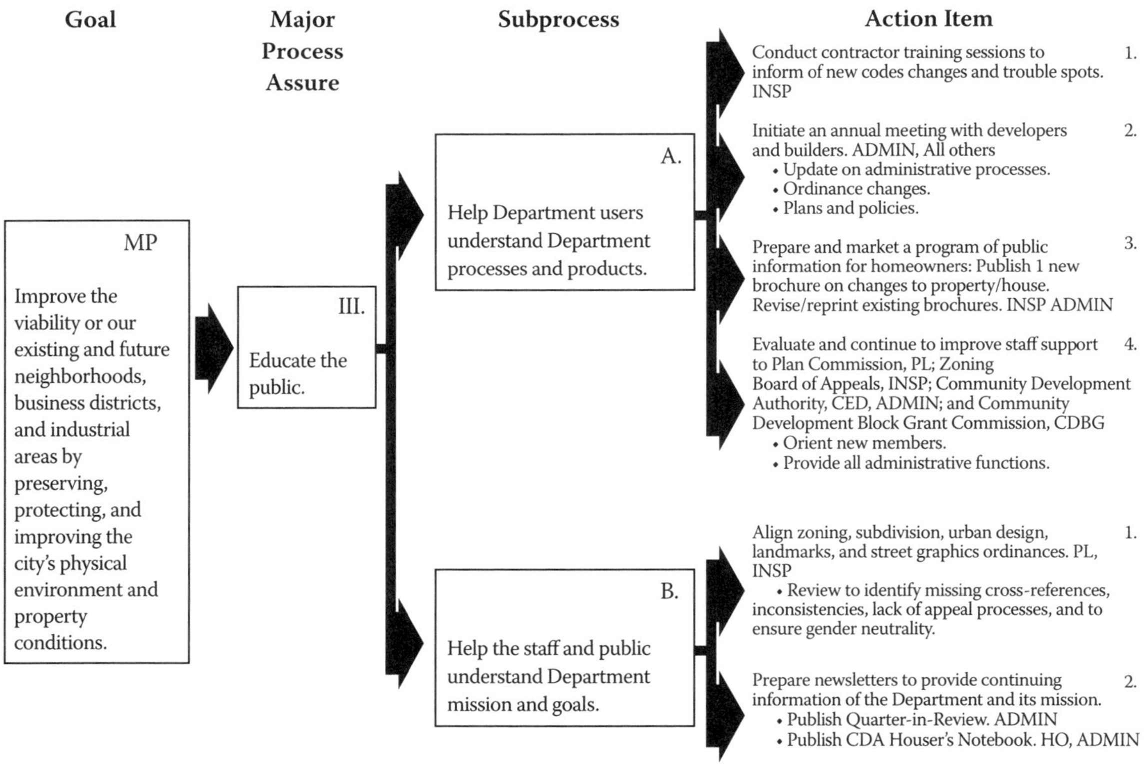

FIGURE 11.1
Transforming goals into action. From Frank Voehl, *Operating System Guidebook* (Milwaukee, WI: ASQ, Quality Press, 2001, p. 6).

1. **Executive Summary:** This is written to the scope and level of content so that an outsider can read the summary and grasp the mission of the organization, its overall major issues and goals, and key strategies to reach the goals.
2. **Authorization:** This page includes all of the necessary signatures from the board of directors (if applicable) and other top management designating that they approve the contents of, and support implementation of, the plan.
3. **Organization Description:** This section describes, for example, the beginnings and history of the organization, its major products and services, highlights and accomplishments during the history of organization, and so on.
4. **Assumptions:** These are the assumptions on which the plan is based.
5. **Mission, Vision, and Value Statements:** These statements describe the strategic philosophy of the organization.
6. **Objectives, Goals, and Strategies:** Lists all of the major objectives, strategic goals, timelines, and associated strategies identified during the Strategic Planning process.
7. **Appendices:** The appendices often include information that is somewhat confidential, detail oriented, and/or tends to change a lot, such as the following:

A. **Action Planning:** Specifies objectives, responsibilities, and timelines for completion of objectives.

B. **Description of the Organizational Master Planning Process:** Describes the process used to develop the plan, who was involved, the number of meetings, any major lessons learned to improve planning the next time around, and so on.

C. **Organizational Master Plan Analysis Data:** Includes information generated during the external analysis (for example, environmental scan) and internal analysis (for example, strengths, weaknesses, opportunities, and threats [SWOT] analysis), and includes listing of strategic issues identified during these analyses.

D. **Goals for Board and Chief Executive Officer:** Goals of the board and CEO should be directly aligned with goals identified during planning process. This appendix will list goals for the board and can include recommendations for redesigning board committees to be associated with organization's goals. This appendix also lists goals for the CEO that can be used (along with the CEO job description) to form the basis for performance evaluations of the CEO.

E. **Budget Planning:** Depicts the resources and funding needed to obtain and use the resources needed to achieve the organization's Master Plan goals. Budgets are often depicted for each year of the term of the organization's Master Plan.
F. **Annual Operating Plan:** Describes the major goals and activities to be accomplished over the coming fiscal year.
G. **Financial Reports:** Includes last year's budget (with estimated expenses and the actual amounts spent), this year's current budget (again, with estimated amounts and actual amounts spent), a balance sheet (or in the case of nonprofits, a statement of financial position), income statement (or in the case of nonprofit, a statement of financial activities).
H. **Monitoring and Evaluation of Plan:** Include criteria for monitoring and evaluation, and the responsibilities and frequencies of monitoring the implementation of the plan.
I. **Communication of Plan:** Describes the actions that will be taken to communicate the plan and/or portions of it, and to whom.

IMPLEMENTATION ISSUES

The Annual Operating Plan uses the activities that are scheduled for the coming year as a major input to the yearly planning and budget cycle. This is the way the organization provides the resources required to implement its Master Plan. When we train organizations in Organizational Alignment and its relationships with Lean and Six Sigma, people often ask: "Which should I do first, Lean or Six Sigma?" While our answer is frequently "it depends," we have found that there is a sequence that often works best. It's Lean, Six Sigma, and process redesign combined (see Figure 11.2).

For changes directed toward expanding markets and growing the organization, there is a natural flow:

- Business strategy will drive changes in process. (Note: Letting process, people, or technology drive strategy leads to customer and employee confusion.)
- Process changes will drive changes in organization and technology (e.g., information systems).

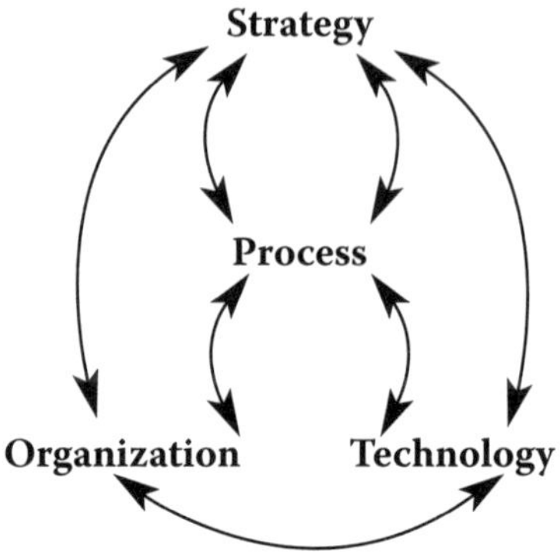

FIGURE 11.2
The interactions among strategy, processes, technology, and the organization.

- Process, organization, and technology changes must be synchronized to maximize the benefits of any change. Failure to keep these in alignment will cause chaos.
- SPOT: The key elements of any transition (see Figure 11.3):
    - **Strategy:** For example, customer satisfaction, innovation, or operational effectiveness
    - **Process:** How things get done
    - **Organization:** Who does what in the process
    - **Technology:** The tools that are used

	A	B	C	D	E	F	G
1	TRANSITION PLAN	FROM →					TO
2		Current Environment	Simplify, Stramline and Standardize (Lean)	Optimize (Six Sigma)	Expand into New Domains (Reengineer)	Competitive Advantage	Becoming World Class
3	STRATEGY (Why?)						
4	PROCESSES (How?)						
5	ORGANIZATION (Who?)						
6	TECHNOLOGY (What?)						

FIGURE 11.3
A SPOT diagram.

Very simply put, Organizational Master Planning identifies where the organization wants to be at some point in the future and how it is going to get there. The *strategic* part of this planning process is the continual attention to current changes in the organization and its external environment, and how this affects the future of the organization. Skills in strategic planning are critical to the long-term success of your organization. This form of planning includes

- Taking a wide look around at what's going on outside the organization and how it might affect the organization (an environmental scan).
- Taking a hard look at what's going on inside the organization, including its strengths, weaknesses, opportunities, and threats.
- Establishing statements of mission, vision, and values.
- Establishing goals to accomplish over the next (usually) five years or so, as a result of what's going on inside and outside the organization.
- Identifying how those goals will be reached (strategies, objectives, responsibilities, and timelines).

SUMMING IT ALL UP

When developing the Organizational Master Plan, you need to bring together all the planning elements. They include the Business Plan and the Strategic Plan, which includes the Strategic Business Plan and the Strategic Improvement Plan, thereby providing a complete plan for the future. You then need to have an Annual Operating Plan that will include the parts of the Strategic Plan that are scheduled to be worked on during the next fiscal year.

> Whenever there is a major change in the Strategic Plan, the organization should consider doing an organizational realignment activity to be sure the organization will be able to implement the new plan.
>
> **—H. James Harrington**

In conclusion, Organizational Master Planning drives and facilitates organizational alignment and determines the overall direction and goals of the enterprise. Consequently, the Organizational Master Plan influences all aspects of the organization, including determining the following:

- The products and services that will be provided by the organization and how those products and services will be designed
- Organizational design and roles that are needed by the organization
- Performance goals that are established for positions throughout the business
- Board committees that should be developed (in the case of corporations)
- Resources that are needed to reach those goals, and consequently, how much money is needed to procure those resources; ultimately, the goals determine the content of various budgets

The two key points to remember while proceeding through this process are

1. The planning process is at least as important as the planning document itself.
2. The planning process is never "done" as the planning process is a series of continuous cycles that are part of the management control process.

Just because you have a compass, it doesn't mean you are going in the right direction.

—H. James Harrington

Too often, an organization devotes a great deal of time, effort, and thought to develop and update the Strategic Plan once a year and then sets it aside until it's dusted off the next year. If this is the case in your organization, don't waste the time and effort. Let the executives play golf instead; it will do more for the organization. Your plan should be a living vibrant thing that is used daily to ensure everyone is marching to the same drummer.

The critical features of a successful Organizational Master Plan include

- Visible leadership commitment
- Regular reporting and sharing of results
- Widespread sharing of the plan throughout the organization
- Well-understood linkage between individual performance and organization's success
- Willingness to change
- Adapting from experience

HP will need to seek growth opportunities with a great deal of insight and skill. We can grow in two ways: by building on capabilities and successes we've already achieved and by innovating new technologies and entering entirely new markets.

—John Young, former CEO, HP

This is HP's Ten-Step Business Plan Methodology:

1. Statement of purpose
2. Five-year objectives
3. Description of customers and channels of distribution
4. Description of competition
5. Description of necessary products and services
6. Plan for development/purchase, and introduction of products and services
7. Financial analysis of costs and returns
8. Potential problems analysis
9. Recommendations
10. First-year tactical plan

When you complete the plan, you are 5 percent of the way to your goal. The real challenge is in implementing it.

—H. James Harrington

The following is a short checklist for the Organizational Master Plan.

- Preliminary tasks
  - Complete market research and analyses.
  - Clarify key matters relating to products/services.
  - Form the basis of the management team.
  - Prepare a list of assumptions that the plan is based upon.
  - Prepare a Strategic Plan as a framework for a detailed plan.
  - Decide the central purpose of the plan and its target audience.
  - Find professional advisers to assist with the planning.
  - Acquire any software tools needed to help prepare the plan.
  - Research and compile a list of possible recipients of the plan.
  - Ascertain any specific needs of likely key recipients.
- Writing the plan
  - Create a framework for the plan, for example, a table of contents.
  - Identify possible appendices, attachments, and so on.

    - Estimate page lengths for each key section.
    - List the main issues and topics to be covered within key sections.
    - Assign work programs based on the framework and lists.
    - Draft all key sections in a logical sequence.
    - Check the preliminary draft for completeness and plug gaps.
    - Stand back and take a detached overview of the draft.
    - Let an outsider or adviser critique the latest draft.
    - Redraft, fine-tune, and check the spelling in the document.
    - Write the executive summary.
    - Develop the plan's conclusion.
    - Get an independent assessment of the final draft.
- Reviewing the plan
    - Is the plan nicely presented—bound, pages numbered, and so on?
    - Has the plan been spell-checked in its final form?
    - Is the plan's length appropriate to its purpose?
    - Have the business's (funding) needs been clearly stated?
    - Does the plan's summary stimulate interest?
    - Have all key questions been anticipated?
    - What likely objections remain unresolved?
    - Will the plan provoke the desired responses?
    - Has our change management process prepared the targeted people to accept the changes that are required?

Why is it that you never have the time to do planning now? It is often because you didn't plan for what you did yesterday.

—H. James Harrington

Appendix A: Definitions and Abbreviations

Annual Operating Plan: A formal statement of business short-range goals, the reasons they are believed to be attainable, the plans for reaching these goals, and the funding approved for each part of the organization (budget). It includes the implementation plan for the coming years (years 1–3) of the Strategic Plan. It may also contain background information about the organization or teams attempting to reach these goals. One of the end results is a performance plan for each manager and employee who will be implementing the plan over the coming year. The Annual Operating Plan is often just referred to as the Operating Plan (OP).

authority: The granting of certain rights to the position holder including the right to give direction to others, and the right to reward and punish. These rights are called *positional powers* and belong to the position rather than to the position holder.

beliefs: An acceptance of something as being true. It, in fact, may or may not be true.

budget: Budgets fund the departmental activities and are used to estimate and plan expenses. Budgets provide the resources required to implement the tactics.

business: The utilization of capital as performance solutions to produce value in the form of results.

Business Plan: A formal statement of a set of business goals, the reasons they are believed to be attainable, and the plan for reaching those goals. It also contains background information about the organization or team attempting to reach those goals.

business process improvement (BPI): Covers the breakthrough improvement approaches process redesign, process reengineering, and process benchmarking.

communication diagram: A pictorial view of the way information is transferred within an organization.

core capabilities: The business processes that visibly provide value to the customer (e.g., Honda's dealer management processes).

core competencies: The technologies and production skills that support an organization's products or services (e.g., Sony's skill at miniaturization).

critical success factors: The key things that the organization must do exceptionally well to overcome today's problems and the roadblocks to meeting the vision statements.

customer-centric organization: An organization whose offerings combine and integrate advice, services, and/or software in support of their products and services, which results in customized offerings. They are driven by customer portfolios.

customer-focused organization: An organization that uses extensive market research in defining their product offerings and design. They build a demand for the products they are able to produce. They invest in providing their frontline employees with the knowledge, products, and tools to provide effective and consistent customer service.

department: A department is made up of a manager and the people who report directly to him or her. This term covers all levels of management—top, middle, and line.

division of labor: The distribution of responsibility (*hierarchy* defines the distribution of authority). Division of labor concerns the way jobs are grouped into organizational units.

executive team: A management team, which usually includes the CEO, COO, and vice presidents of an organization. In some organizations, senior key staff people are also included.

external supplier: A supplier that is not part of the customer's organizational structure.

Fast Analysis Solution Technique (FAST): A breakthrough approach that focuses a group's attention on a subprocess for a one- or two-day meeting to define how the group can improve the process over the next 90 days. Before the end of the meeting, management approves or rejects the proposed improvements.

flowchart: A method of graphically describing a process (existing or proposed) by simple symbols and words to display the sequence of activities in the process.

function: An organization that is made up people with a specific skill set (i.e., Quality Assurance, Finance, Human Relations, Production Control, Sales, Marketing, Product Engineering, Manufacturing, Industrial Engineering, etc.) or the total group of people that reports directly to a vice president of the organization.

hierarchy: Reflects the distribution of authority among organizational positions. It defines formal reporting relationships that map upward communication channels through which management expects information to flow.

High-Impact Team (HIT): A breakthrough approach that focuses a group's attention on the processes that are going on within a specific area. It realigns the work area to minimize the movement of output between activities resulting in decreasing stock and shorter cycle time. A typical HIT activity will last for two weeks, and between 70% and 80% of the future-state solutions will be implemented within the two-week time period.

individual performance indicators (IPI): A set of measurements that is used to measure the performance of individual managers and employees. These are key inputs to their performance plan and evaluation. Often they are included in the individual's job description.

internal supplier: An area within an organizational structure that provides input into another area within the same organizational structure.

key business driver (KBD): Things within the organization that management can change that control or influence the organization's culture and the way the organization operates (also called *controllable factors*).

Key Business Driver Improvement Plan: Plans designed to transform the organization from the present state so that it is in line with the relevant KBD vision statement. They are usually 3- to 5-year plans.

key business driver (KBD) vision statement: Statements of how the key business drivers will be operating 5 years in the future. They are developed during the strategic improvement planning process. There are usually 8 to 12 of these vision statements (also called controllable factor vision statements).

key performance indicator (KPI): A set of measurements, usually no more than ten, which are used to evaluate the progress and performance of a total organization.

long-term vision statement: Usually prepared by top management; defines the future state of an organization 10 to 25 years in the future from the date that the statement is created.

mission: The stated reason for the existence of the organization. It is usually prepared by the chief executive officer and seldom changes, normally only when the organization decides to pursue a completely new market.

natural work team: Any group of people that consists of a manager or a team leader and the people who report directly to him or her.

organization segment: Smaller subgroups comprising like or supporting types of activities. They may be divided by geographic, industrial/market segment, or user client needs.

Organizational Master Plan: The combination and alignment of an organization's Business Plan, Strategic Business Plan, Strategic Improvement Plan, Strategic Plan, and Annual Operating Plan.

organizational alignment: The methodology that brings the organization's structure, processes, networks, people, and reward system in harmony with the Strategic Business Plan and the Strategic Improvement Plan.

organizational objective: Used to define what the organization wishes to accomplish over the next 5 to 10 years. These are usually defined by top management.

outcome: The measured results that the organization realized as a result of the action taken.

performance goal: Quantifies the results that will be obtained if the organizational objectives are reached.

performance plans: Contracts between management and the employees that define the employees' roles in accomplishing the tactics, and the budget limitations that the employees have placed upon them.

planning assumptions: The conditions and inputs that the individual, who is doing the planning, will consider as part of his/her estimation. They are things that the planner believes will happen during the planning period. (Example: Productivity will increase by 10% or we will start shipping the new product on July 15.)

planning ground rules: The conditions and requirements on which the plan is based. The organization's current conditions are a key input to the planning ground rules. The requirements include setting targets for the organization's KPIs.

policy deployment: A method devised to capture and cement strategic goals as well as flashes of insight about the future and develop a means to bring these into reality. It is based upon Shewhart's plan-do-check-act cycle. It creates goals, selects control points, and links daily control activities to the organization's strategy. Also called *hoshin kanri*.

principle: The ultimate source, a fundamental truth, the motivating source upon which others are based. If a statement is ever not true or not followed, it is not a principle. (Note: Any manager or employee who is not living up to the organization's principles is a good candidate for separation from the organization.)

process map: A hierarchical method for displaying processes that illustrates how a product or transaction is processed. It is a visual representation of the workflow within a process, or as an image of the whole operation. Process mapping comprises a stream of activities that transforms a well-defined input or set of inputs into a predefined set of outputs. It is a flowchart with inputs and outputs added to each activity thereby increasing its value in refining processes.

product-centric organization: An organization that has multiple product lines that divide into separate business lines and/or models. Often there are few or no interrelationships between the product lines (i.e., a computer manufacturer that sells management consultant services also). They are driven by product portfolios.

short-term vision statement: Usually prepared by top management and defines the desired future state of an organization 5 or 10 years in the future from the date that the statement is created.

Strategic Business Plan (SBP): Focuses on what the organization is going to do to grow its market. It is designed to answer the questions: What do we do? How can we beat or avoid the competition? It is directed at the product and/or services that the organization provides as viewed by the outside world. When an organization is just being funded, this plan is often just called a Business Plan.

strategic excellence positions: Unique and distinctive capabilities that are valued by the customer and provide a basis for competitive advantage (e.g., Avon's distribution system).

strategic focus: Key things that will set the organization apart from its competitors over the next 5 years. This list is defined by top and middle-level managers.

Strategic Improvement Plan (SIP): Focuses on how to change the culture of the organization. It is designed to answer the questions: How do we excel? How can we increase value to all the stakeholders? It addresses how the controllable factors within the organization can be changed to improve the organization's reputation and performance.

strategic management: The process of specifying the organization's mission, vision, and objectives, developing policies and plans, often in terms of projects and programs that are designed to achieve these objectives and then allocating resources to implementation of policies and plans, projects and programs.

Strategic Plan: A document that is the result of strategic planning. It defines the organization's strategy and direction and makes decisions on the allocation of resources in pursuit of the organization's strategy including its capital and people. It focuses on the future of the organization and is the combination of the Strategic Business Plan and the Strategic Improvement Plan, with each item prioritized to maximize the organization's performance.

strategic planning: The systematic and more or less formalized effort of an organization to establish basic organizational purposes, objectives, policies, and strategies and to develop detailed plans to implement policies and strategies to achieve objectives and basic organizational purposes.

strategy: Method that defines the approach that will be used to meet the performance goals.

supplier: An organization that provides a product (input) to the customer.

supplier controls: The preventive measures that are put into place to minimize the possibility of suppliers providing unacceptable product. They include things like supplier qualification, requirements placed on the supplier to be ISO 9000 certified, source inspection, and receiving inspection,

supply chain management: The creation of a management process for integrating decisions, plans, and information systems from customer requirements through the fulfillment process to the suppliers of materials.

tactic: Method of defining how the strategies will be implemented. Tactics explain how the strategies will be accomplished.

transition: An orderly passage from one state, condition, or action to another.

value: Things that an organization holds as sacred; values should never be sacrificed. Much care should be taken when an organization selects its values to be sure they do not conflict with each other. Values are often interchanged with beliefs, but management is held accountable for complying with values.

value statements: Documented directives that set behavioral patterns for all employees. These are the basic beliefs that the organization is founded upon, the principles that make up the organization's culture. Values are deeply engrained operating rules or guiding principles of an organization that should not be compromised; they are rarely changed. Values are sometimes called operating principles, guiding principles, or basic belief operating rules.

vision: A documented or mental description or picture of a desired future state of an organization, process, team, key business driver, or activity.

vision statements: Prepared by top management and directed at what the organization's output will be like and/or how it will be used 10–25 years in the future. Some organizations have a long-term vision statement (10–25 years) and a short-term vision statement (5–10 years).

Appendix B: Improvement Tools

The following is an alphabetized list of some of the more than 1,100 improvement tools and methodologies:

7-S Model (Organizational Effectiveness)
Acceptable Process Level
Acceptable Quality Level
Acceptance Control Charts
Acceptance Sampling
Acquisition Streamlining
Action Diagramming
Action Planning
Activity Accounting
Activity Analysis
Activity Based Costing
Activity Cost Pool Definition
Affinity Diagrams
Analysis and Segmentation of Customer Views
Analysis of Customer Wants
Analysis of Variance/ANOVA
Annual Strategic Quality Plans
Application Construction
Application Installation
Application Structuring and Identification
Application Testing
Area Activity Analysis (AAA)
Association Diagramming
Assumptions Evaluation
Attribute Acceptance Sampling
Attribute Control Chart
Attribute Data
Attribute Identification
Attribute Measurement
Attribute Sampling
Attribute Sampling Procedures

Attribute Sampling Tables
Auditing
Audits by Top Management
Automatic Test Equipment
Automation
Autonomation
Average
Average (x) Chart
Awards
Baldrige Award
Barplots
Batch Procedure Design
Bayes' Theorem
Benchmarking
Benefits Assessment
Big-Picture Analysis
Black Belt Training
Block (random) Sampling
Box Plots
BPI Measurement Methods
Brainstorming
Bureaucracy Elimination Methods
Business Area Data Modeling
Business Performance Management (BPM), BPM Measurement
Business Strategy Analysis
Business Systems Planning
Business Transaction Identification
C Charts
Capacity and Staff Planning
Career Development
Career Planning
Cash Bonuses
Cause-and-Effect Diagrams (Fishbone Diagrams)
Cause-and-Effect Diagrams with Cards (CEDAC)
Central Tendency Measurement
Chain Sampling Plans
Charts
Checklists
Checksheet Design

Chi Square Distribution/Test
Classification of Characteristics
Clearance Fits
Clearing Interval
Code Generation
Competitive Analysis
Competitor Product Disassembly Research
Computer Simulation
Computers
Concurrent Engineering
Conditional Probability
Consensus Building
Consumer Risk Quality (CRQ)
Context Diagramming
Continuous Sampling Plan
Contract Negotiation
Control Array
Control Charts: c Charts, P Charts, u Charts, X Bar and R Charts
Control Limits
Control of Quality
Corrective Action
Correlation Analysis
Cost Effectiveness Programs
Cost Flow Diagramming
Cost of (Poor) Quality (COPQ)
Cost/Benefit Analysis
Cost-Cycle Time Analysis
Cost-Driven Analysis
Cost–Time Analysis
Cost–Time Charts
Could Cost
Counting Rules
Creative Brainstorming
Critical Success Factor Analysis
Criticality Analysis
Critical-to-Quality (CTQ)
Cross-Functional Management
Crossplots
Cumulative Hazard Sheet

Cumulative Sum Control Chart
Current Systems Investigation
Customer Analysis
Customer Data Analysis and Action Plans
Customer Interface Training
Customer Loss Analysis (Cost Impact)
Customer Partnerships
Customer Phone Calls (Management and Employees)
Customer Requirements Mapping
Customer Reviews
Customer Roundtables
Customer Satisfaction Standards/Ratings
Customer Simulated Testing
Customer Surveys
Customer Visits
Customer-Related Measurements
CUSUM Chart
Cycle Time Reporting
Cycle-Time Reduction Methods
Data Access Modeling
Data Collection
Data Flow Diagramming
Data Gathering by Document Review
Data Gathering by Interview
Data Gathering by Samples and Surveys
Data Gathering by Secondary Research
Data Sheets
Data Stratification
Data Structure Diagramming
Data View Identification
Database Generation
Decision Limits
Delphi Methods
Delphi Narrowing Technique
Department Improvement Teams (DIT)
Design of Experiments
Design Quality
Design Review
Design-to-Production Transition

Discreet Event Simulation
Discreet Probability Distributions

- Binomial
- Discreet Uniform
- Hypergeometric
- Multinomial
- Negative Binomial
- Poisson

Dispersion Measures
Distributions; *see also* normal distribution; probability distribution; reliability; chi-square
Documentation Design
Dodge-Romig System
Double Sampling Plans
Double Specification Limit Plans
Education Cost Sharing
Education Design and Development
Effective Delegation
Effective Listening
Empirical Quantile Plots
Employee (Satisfaction) Surveys
Employee Training
Employee Visits
Environmental Pride Process (Your Living Rooms)
Error-Proofing
Errorproofing Fixtures and Methods
Established Reliability
Estimation
European Quality Award
EVOP
Executive Improvement Teams (EIT)
Executive Information Needs Analysis
Executive Needs Analysis
Exemplary Facilities
Expected Values
Exponential Formula for Reliability; Exponentially Weighted Moving Average; EWMA Chart
F Distributions
Facilitator Training

Facilitators
Facilities Planning
Failure Analysis
Failure Mode Effect Analysis
Fault Tree Analysis
Failure Mode Effects Analysis (FMEA)
Fewer Good Suppliers
Field Reporting
Financial Analysis
Financial Reporting
Five Whys
Flattening the Organization
Flowcharting
FOCUS
Focus Groups
Foolproof Engineering Methods
Force Field Analysis
Forms Design
Fractional Acceptance Numbers Plan
Frequency Distributions
Function Diagrams
Gain Sharing
General Ledger Analysis and Consolidation
Geometric Distributions
Goal Control Process
Good Manufacturing Practices (GMPs); *see also* CGMPs
Graphical Methods
Group Presentation
Group Recognition
Group Sequential Plans
Harrington's Ten Fundamental Tools

- Bureaucracy
- Business Process Improvement Concepts
- Cost and Cycle Time Estimating (Activity-Based Costing)
- Customer Needs/Expectation Assessment
- Flowcharting
- Measurement Methods
- Non-Value-Added Activities Elimination
- Process Simplification

- Simple English
- Walk-Throughs

Harrington's Ten Sophisticated Tools

- Benchmarking
- Business Systems Planning (BSP)
- Information Engineering
- PERT Charting
- Poor-Quality Cost
- Process Analysis Technique (PAT)
- Quality Function Deployment (QFD)
- Structured Analysis and Structured Design (SASD)
- Value Analysis
- Value Control

Hawthorne Effect
Histograms
Historical Analysis
Hoshin Planning
Human Factors Engineering
Hypergeometric Distribution
Hypothesis Testing
Improvement Process Measurements
Indifference Quality Level
Individual Private Recognition
Individual Public Recognition
Individuals (x) Chart
Industrial Modernization Incentives Program (IMIP)
Inference Testing: see Statistical Inference
Information Engineering
In-Process Data Collection
Inspection and Testing
Inspection Planning
Instruments
Interference Fits
Internal Customer Concepts
Internal Customer Measurements
Interviewing Techniques
Investment Management
Ishiwaka Diagram; *see also* Ishiwaka Fisbone; Fishbone Cause-and-Effect Diagram

ISO-9000 Compliance; ISO-2959-1; ISO-3534-2; ISO-3951; ISO-8402; ISO-19011
Job Descriptions
Job Training and Certification
Job-Related Training
JUSE
Just-In-Time
Kano Method; Total Quality Creation
Kolmogorov-Smirnov Test
Kume Quality Rules
Leadership Skills Development
- Coaching
- Communication
- Inspiring
- Listening
- Modeling

Leadership Skills Training
Life Cycle Costing
Limiting Quality; Limiting Quality Level
Line Management
Location Plots
Logical Database Design
Lognormal Distribution
Long-Range Quality Planning
Loss Function; *see also* Taguchi Loss Function
Lot Plot
Lot Tolerance Percent Defective (LTPD)
LQL; *see* Limiting Quality Level
Maintainability Analysis
Maintainability Assessment
Management by Objectives
Management by Objectives Achievement Measurement
Management by Walking Around
Management Improvement Teams (MIT)
Management Presentations
Management Self-Audits
Management's Seven Tools
- Affinity Program (K-J)
- Arrow Diagram

- Interrelationship Diagrams
- Matrix Charts
- Matrix Data Analysis
- Process Decision Program Chart
- Tree Diagram

Manufacturability Assessment
Market Analysis
Matched Plans, derivation of
Maximum Standard Deviation
Mean Time Between Failures (MTBF)
Measurement; central tendency, robust design,
Median Range
Median x Chart
Meeting Management
Metrology
MIL-Q-9858A Compliance; MIL-HDBK-217,338; MIL-STD-105D, 105E, 756, 756B, 781, 785, 1472, 1543, 1629/35
Mind Maps
Mission Statements
Modified Acceptance Control Charts, Modified Control Charts, Modified Control Limits
Monetary Awards
Monte Carlo Sampling
Moving Range
Multiple Sampling Plan
Multiple-Attribute Decision Modeling (MADM)
Multi-Skills Maintenance
Multi-Skills Operator
Multi-Vari Strategy
Multivariable Analysis; Multivariate Analysis
Narrow-Limit Gauging
National Association of Testing Authorities (NATA)
Negative Analysis
New Employee Selection
New Performance Standards
Noise Plan, Noise variables, Nomograph Continuous Sampling Plans (CSP)
Nominal Group Techniques
Non-Conformances, Non-Conforming Spacing
Nondestructive Testing and Training

Non-Verbal Communications
Normal Distribution
Notched Box Plots
No-Value-Added Analysis
np Charts
On-Line Conversion Design
Operating Characteristic (OC) Curve
Operation Verification
Operational Alternatives Analysis
Operational Definitions
Organization Design
Organizational Analysis
Organizational Change Management Techniques
Organizational/Human Resource Enablement Sessions
Orthogonal Polynomial
Package Software Evaluation
Package Validation Testing
Paperwork Simplification Techniques
Pareto Diagram; *see also* the 80/20 Rule = the Pareto Principle; the Pareto Pyramid
Participative Management
Pay for Knowledge System
Pay for Performance
Pearson Coefficient of Correlation
Performance Planning & Evaluation
Personality Profile (Keirsey-Bates)
Physical Database Design
PIC-A-Solution
Plan, Do, Check, Act Cycle
Planned Experimentation, robust design
Poisson Distribution
Policy Deployment; *see also* Hoshin Kanri
Poor-Quality Cost; *see also* Cost of Quality
PRE-Control
Principal Component
Principal Component Analysis
Prioritization Matrixes
Prioritization through Ratings
Probability Distributions

Problem Solving (see II)
Problem Tracking Logs
Procedure Identification and Specification
Process Analysis and Improvement
Process Analysis Techniques
Process Capability Analysis
Process Capability Studies
Process Control Techniques
Process Decision Program Chart
Process Documentation
Process Engineering
Process Flow Controls
Process Improvement Teams (PIT)
Process Modeling
Process Qualification
Process Performance and Capability
Process Simplification Techniques
Process Walk-Through Methods
Process Window Definitions
Producers Risk Quality (PRQ)
Product Cycle Controls
Product Design Assurance
Product Quality, eight dimensions of
Program Evaluation and Review Technique (PERT) Charting
Projection Analysis
Prototype Test Checklists
Prototyping
Quality, definition of; auditors; control of; design vs. manufactured; specifications/requirements for
Quality Area Improvement
Quality Assurance Planning
Quality Characteristics
Quality Communication
Quality Company Policies
Quality Control, *see also* Control of Quality
Quality Control Circles (QCC)
Quality Data Collection and Reporting System
Quality Engineering Methods and Training
Quality Function Deployment

Quality Improvement; also Quality Improvement Teams (QITs)
Quality Integration
Quality Loss Function
Quality Manuals
Quality Policy
Quality Policy Deployment
Quality Systems
Quality Visions
Quantile–Quantile Plots
Quincunx; Quincunx Data
Random (Block) Sampling
Range (R) Chart
Regression Analysis
Reject Control Charts
Rejectable Process Level (RPL)
Reliability Analysis
Reliability Predictions
Reliability-Centered Maintenance
Report Design
Requests for Corrective Action (RCA)
Requirements; *see also* Customer Requirements
Resource and Activity Driver Analysis
Responsibility Charting
Restructuring of the Quality Assurance Organization
Reverse Thinking
Risk Assessment
Risk Taking
Risk/Opportunity Management Process
Robust Design
Rules of Probability, Combinatorics
Sample; Sample Averages/Range; Sample Plan; Sampling System/Scheme
Sample Inspection
Sampling Techniques; *see also* Acceptance Sampling
SCAMPER (Substitute/Combine/Adapt/Magnify/Put to other uses/Eliminate/Range)
Scatter Diagram; Scatter Plots
Screen Design
Security and Access Control Design
Self-Control; Process Control Techniques

Self-Managed Work Teams
Sequential Analysis; *see also* Sequential Probability Ratio
Sequential Sampling Plans
Set-Up Time Reduction
Seven Basic Tools

- Brainstorming
- Cause and Effect Diagrams
- Check Sheets
- Data Stratification
- Histograms
- Pareto Diagrams
- Scatter Diagrams

Ship-to Stock Cost
Simple English
Simple Language Analysis
Single Specification Limit Plans
Six Sigma; system of measurement and problem solving
Skip Lot Inspection; Skip-Lot Sampling Plans
Software Quality Assurance
Solution Analysis Diagrams
Solutions Evaluation
Source Inspection
Specification; Specification Limits and Tolerances
Stakeholder Needs Analysis
Standard Deviation; Charts/Method
Standards; *see also* Quality Standard
Standardization
Statistical Design of Experiments

- Factorial Design
- Fractional Factorial Experiments
- Grace Latin Square
- Latin Square Model
- Orthogonal Arrays

Statistical Estimation

- Bayesian Estimates
- Confidence Intervals
- Statistical Inference
- Prediction Intervals
- Tolerance Intervals

Statistical Methods (Control Charts)
Statistical Process Control

- c Charts
- np Charts
- p Charts
- u Charts
- X Bar and R Charts

Statistical Thinking
Statistical Tolerance Intervals/Levels/Limits
Stem and Leaf Plots
Stock Purchase Plans
Storyboarding
Strategic Business Review
Strategic Planning Process; *see also* Annual Business Planning; Operations Planning
Stress Testing
Stress Testing (Test to Failure)
Structural Customer Surveys
Structural Methodology for Process Improvement
Structure Charts
Structured Analysis
Structured Customer Surveys
Structured Design
Substitute/Combine/Adapt/Magnify/Put to other uses/Eliminate/Range (SCAMPER)
Suggestion Programs
Supplier Design Involvement
Supplier Partnerships
Supplier Process Audits
Supplier Qualification
Supplier Quality Incentive Plans
Supplier Ratings
Supplier Seminars
Supplier Surveys
System Reliability
Systematic Sampling Plan
Systems Assurance
Taguchi Techniques
Target Costing

Target Goal Setting
Task Team (TT)
Team Building
Team Recognition
Teams-Group Process
Technology Enablement Sessions
Technology Impact Analysis
Test Objectives Definition; Test Equipment
Test of Hypotheses
Test Plan Design
Theoretical Quantile–Quantile Plots
Three Dimensional Bar Plots
Time Management
Tolerance Limits; Tolerances
Total Quality Management (TQM)
Total Strategic Quality
Training Programs
Type I, II, III Conflicts
Value Analysis
Value Analysis Engineering
Value Engineering
Variability
Variable Control Charts
Variables data; *see also* Variables Acceptance Plan
Variation
Visioning; also Vision Statements
Visual Controls
V-Mask
White Collar Production Management
Wiebull Chart
Work Cells
Work Flow Analysis
Work Simplification
Work Teams
X Charts
Zero Defects
Zero Stock (Just-In-Time)

Appendix C: Problem Analysis Cycle

Everyone flowcharts their processes, but once this is done, the real challenge begins. The question is, "How can you streamline the process?" The following is one simple approach that has worked for me:

Step 1: Review each block in the flowchart to define the classification into which each block.

- **Real-value-added activity (RVA):** These are activities that are directly related to the product or service that will be delivered to the external customer, and are activities for which the customer would be willing to pay. (*Example:* machining a part, cooking a meal in a restaurant, writing out a sales order. etc.)
- **Business-value-added activity (BVA):** These are activities that are needed to run the organization but not things that the external customers want you to do for them. (*Example:* preparing budgets, filling out employee records, updating operating procedures, etc.)
- **No-value-added activity (NVA):** These are activities that aren't necessary. (*Example:* inspecting parts, checking the accuracy of reports, reworking a unit, rewriting a report, etc.)

Figure C.1 is a simple four-block flowchart. The first operation is machining a part, which is an *RVA* activity; the second operation is recording the time that it takes to machine the part, which is a *BVA* activity; the third operation is inspecting the part to see if it is good or bad, which is an *NVA* activity.

To highlight these differences, we recommend coloring in the RVA activities in green, the BVA activities in yellow, and the NVA activities in red. This is called a *rainbow flowchart.* Often, bureaucracy-type NVA activities are colored in blue to set them apart.

Step 2: Define the NVA activities that are inspection or audit activities.

Step 3: For the inspection and audit activities, define what they cost and how much they delay the process. Next, compare them to the RVA content that the inspection or audit adds because it keeps defective

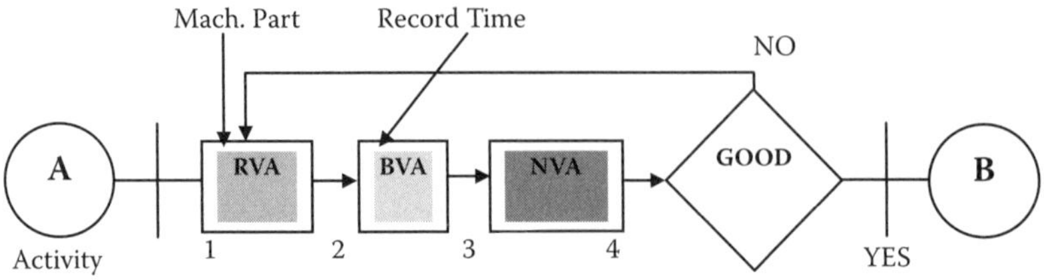

FIGURE C.1
Process flow diagram.

items from moving on to later higher-cost activities. If the RVA content is less than the cost of doing the audit or inspection, consider eliminating the activity. One word of caution: often, people do a better job because they know that their output will be reviewed or inspected by another person, and if the inspection or audit is dropped, the quality of the work may become unacceptable.

Step 4: For those inspection and audit operations that will remain, collect the defect, error, or deviation data and plot them using a Pareto diagram (see Table C.1 and Figure C.2). For the top 50 to 60% of the defects, use the "5 Whys" technique to define the root causes. In some cases, a more sophisticated root-cause analysis maybe required.

Step 5: Perform a root-cause analysis on the three highest defects: B, E, and F. Table C.2 shows the results.

For defect B, two root causes were defined (BR1 and BR2) and only one root cause was defined for defects E and F (ER & FR).

Step 6: Now, for each of the root causes, a corrective action plan needs to be developed (see Table C.3).

TABLE C.1
Number of Defects by Defect Name

Defect Name	Quality
A	3
B	12
C	2
D	0
E	8
F	10
Miscellaneous	7

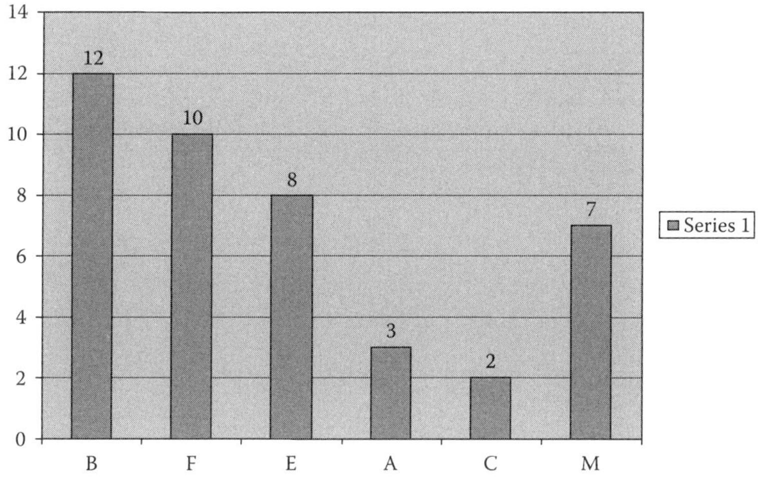

FIGURE C.2
Pareto diagram of defects.

TABLE C.2
Root Cause per Defect

Defect	Root Cause
B	BR1
	BR2
E	ER
F	FR

TABLE C.3
Root Causes and Their Related Corrective Action

Root Cause	Corrective Action
BR1	BR1C1
BR2	BR2C1
	BR2C2
	BR2C3
ER	ERC1
	FRC2
	ERC3
FR	FRC1
	FRC2

TABLE C.4

Impact of Corrective Action

Corrective Action	Saved ($)	Reduced Variation	Quality Improvement	Cycle Time	Inventory Reduction	Customer Satisfaction	Processing Time
BR1C1							
BR2C1							
BR2C2							
BR2C3							
ERC1							
ERC2							
ERC3							
FRC1							
FRC2							

Step 7: We will now analyze each corrective action to define what impact it would have on the process/organization. Each corrective action is evaluated as a stand-alone item (see Table C.4).

It is important to understand that a single corrective action can have impact in a number of ways. For example: by reducing variation in a process, process cycle time and cost can be decreased and quality and customer satisfaction can be improved. As a result, in Table C.4 we analyze each corrective action in the following ways:

- Dollars saved
- Reduced variation
- Quality improvement
- Cycle time reduction
- Inventory reduction
- Increased customer satisfaction
- Processing time

Step 8: The next activity is to study the impact each of the corrective actions has on the different defects (see Table C.5).

Step 9: For each corrective action, evaluate its cost, cycle time, work load, and difficulty to implement (see Table C.6)

Step 10: By analyzing Tables C.4, C.5, and C.6, define the sequence in which the improvements will be implemented. As you define the implementation order, add the improvement to the impact table (see Table C.7).

TABLE C.5

Corrective Action Impact on Defects

	Defect Types						
CA	**A**	**B**	**C**	**D**	**E**	**F**	**Misc.**
BR1C1	+	++	–	0	+	+	0
BR2C1	+	++	+	+	0	0	0
BR2C2	–	++	0	0	+	0	+
BR2C3	0	+	0	0	0	0	0
ERC1	0	X	–	0	++	0	0
ERC2	0	+	0	0	++	+	0
ERC3	0	0	0	0	++	0	0
FRC1	0	+	0	0	+	++	+
FRC2	+	0	0	–	–	+	0

Note: + Positive, ++ Very Positive, 0 No Impact, – Negative, –– Very Negative

TABLE C.6

Corrective Action Implementation Analysis

	Cost ($)	**Cycle Time**	**Difficulty**	**Work Load**
BR1C1	10,000	30 days	Medium	5 days
BR2C1	2,000	10 days	Low	1 day
BR2C2	25,000	45 days	Medium	30 days
BR2C3	3,000	5 days	Medium	6 days
ERC1	1,000	4 days	Medium	1 day
ERC2	18,000	18 days	Medium	10 days
ERC3	28,000	30 days	High	60 days
FRC1	5,000	3 days	Low	1 day
FRC2	15,000	10 days	Medium	8 days

TABLE C.7

Implementation Impact

	% Defects Reduction									
CA	**A**	**B**	**C**	**D**	**E**	**F**	**M**	**Savings ($)**	**Cycle Time**	**Customer Satisfaction**
BR2C1	33	40	10	0	0	0	10	25,000	–10 hrs	+
FRC1	0	30	0	0	15	75	10	30,000	0	+
BR1C1	33	30	–10	0	20	25	0	1,000	–5 hrs	+
ERC2	0	0	0	0	50	0	20	1,000	0	0

Note: CA = corrective action.

In Table C.7, we recorded the percentage of the defect problems corrected by the corrective action. Some people prefer to analyze the number of defects that would be eliminated by each corrective action instead of the percentage of the problem that would be eliminated. For example: Corrective Action BR2C1 would reduce the 3 defects in Figure C.2 Problem A by one defect instead of 33%. Both approaches are acceptable.

Each corrective action must assume that the problem has been reduced as defined by the previous corrective action. Example: If the previous corrective action eliminated the burs on a part by using a new tool, a later corrective action could not claim that it had eliminated the burs by changing the materials.

In analyzing the projected impact of improvement for the first three corrective actions, you will note the following results: defects for Problem B are completely eliminated; defects for Problem A are reduced by 66%, Problem F is solved, and defects for Problem E are decreased by 35%.

In the Problem E example, after the first three corrective actions were selected and the impacts analyzed, additional improvement was still needed. So the fourth corrective action that was planned for implementation was ERC2. The combination of these four corrective actions was enough to correct the process, and five of the potential corrective actions did not need to be implemented.

SUMMARY

This approach is a simple and straightforward way of completing a thorough problem-analysis cycle. The first step defines the root causes and the next steps analyze the potential corrective actions to optimize the positive effects of implementing those corrective actions on the total process.

Index

P

Q